Cognitive Systems Monographs

Volume 42

The Cognitive Systems Monographs (COSMOS) publish new developments and advances in the fields of cognitive systems research, rapidly and informally but with a high quality. The intent is to bridge cognitive brain science and biology with engineering disciplines. It covers all the technical contents, applications, and multidisciplinary aspects of cognitive systems, such as Bionics, System Analysis, System Modelling, System Design, Human Motion Understanding, Human Activity Understanding, Learning of Behaviour, Man-Machine Interaction, Smart and Cognitive Environments, Human and Computer Vision, Neuroinformatics, Humanoids, Biologically motivated systems and artefacts, Autonomous Systems, Linguistics, Sports Engineering, Computational Intelligence, Biosignal Processing, or Cognitive Materials—as well as the methodologies behind them. Within the scope of the series are monographs, lecture notes, selected contributions from specialized conferences and workshops, as well as selected Ph.D. theses.

Indexed by SCOPUS, DBLP, zbMATH, SCImago.

More information about this series at http://www.springer.com/series/8354

Joachim Diederich

The Psychology of Artificial Superintelligence

Springer

Joachim Diederich
The University of Queensland
School of Information Technology and Electrical
Engineering
St Lucia, QLD, Australia

ISSN 1867-4925 ISSN 1867-4933 (electronic)
Cognitive Systems Monographs
ISBN 978-3-030-71841-1 ISBN 978-3-030-71842-8 (eBook)
https://doi.org/10.1007/978-3-030-71842-8

This Springer imprint is published by the registered company Springer Nature Switzerland AG
The registered company address is: Gewerbestrasse 11, 6330 Cham, Switzerland

For Susan

Preface

This book explores the psychological impact of advanced forms of artificial intelligence. There are psychological consequences for the well-being of individuals as well as a significant impact on societies. Human work will continue to be transformed and will possibly be eliminated in the not-so-distant future. Interfaces that directly connect the brain with the internet will have an impact on how we think and communicate. The decisions and actions of an advanced form of artificial intelligence will be more and more difficult to understand, and hence, better forms of explanation for artificial intelligence are required. The technology is increasingly being used to manage significant parts of society, e.g. by use of social credit systems, with consequences for the entire population. Finally, advancements in military AI may include autonomous killing machines that can spread fear and terror. All these developments are happening as we speak and represent significant challenges to human psychological well-being.

Clearly, there are advancements in the medical sector through improved artificial intelligence, and human lives may be enriched in other areas as well. But there needs to be an informed discussion about the risks and challenges of such advancements. There needs to be information so that individuals can make a decision on the use of the technology or the exposure to new developments.

The core message of this book is that advanced forms of artificial intelligence will have an impact on everybody: The developers and users of AI systems as well as individuals who have no direct contact with this form of technology. This is due to the *soliciting* nature of artificial intelligence. The *universal solicitation* of the technology is the challenge. A solicitation that can become a demand.

Machines "want" to be used because they directly relate to human needs. This goes beyond the concept of affordance since the form or shape of the machine must not directly relate to the human requirement. A calculator wants to be used because most humans find it way too time-consuming and exhausting to calculate in the head or to use pen and paper. The GPS in the car wants to be used because people are just about to loose the ability to navigate roads. The heater wants to be used when it is cold. Machines directly address human needs and frequently more than one. A car allows for transport from point a to b but may also represent social status. In addition, a car wants to be used or it breaks down (battery, etc). The artificial superintelligence *wants to be used all the time* because it simplifies and organises human lives, reduces efforts and satisfies human needs. This is called universal solicitation.

By way of example, a military commander in a combat situation may step away from the use of autonomous, weaponised drones guided by artificial intelligence, nevertheless, the knowledge of the technology, its availability and impact will influence future decisions. A person may not actually buy or use a sex robot, however, knowledge of the existence of the devices, imagery associated with the machines and their availability will have an impact on relationships. If someone chooses to live a simple life in a remote area without any high tech, then this choice is the result of the existence of the technology, even if the recluse has no direct exposure to it. There is still the solicitation.

The vision here is a form of advanced artificial intelligence that is not just "human controlled" but beneficial in the sense that it explains itself and its operation to everybody. This includes the most vulnerable in a society, including children, the elderly and persons with an intellectual disability. These explanations must be in a form suitable and acceptable to users. Most probably, these explanations will take the form of videos and demonstrations. Even advanced forms of AI needs to match human comprehension.

Given significant advancements in brain-computer interfaces and the prospect of a direct link between neural and electronic information processing, a number of principles are important from a psychological point of view. The human motivational system must remain operational, in particular the reward system. Motivation breaks down if rewards are delivered to the human brain directly with no action or effort required to gain these rewards. This imposes restrictions on brain-machine interfaces. When the enhancement of cognitive abilities becomes possible by technical means, and at a stage when rewards are delivered directly into the human brain, the risk is that we "do not have to do anything anymore to get something". This is certainly a challenge to human existence.

For those who are not convinced by the concepts of a human controllable AI, there must be the right to withdraw and to live a simpler life. Currently, the Amish people in the USA and Canada refuse the use of advanced technology and live compatibly with the societies surrounding these communities. There needs to be an opportunity to live away from artificial intelligence while still having the option to benefit from some aspects. Many people may seek a simpler life but who does not want to take advantage of progress in healthcare and the delivery of medical services?

The rapid development of artificial intelligence requires a very complex set of decisions for many people; for those who develop and use the technology but also for those who prefer to remain untouched. These decisions are deeply personal in nature. This book aims to provide some of the information required for this decision-making process.

I would like to thank my wife Susan Kay Wright for her love and support as well as significant feedback on early drafts of this book. Luke Diederich, Leonie Holthaus and Peter Trawny provided valuable comments and suggestions.

Queensland, Australia Joachim Diederich

Contents

Chapter 1
Universal Solicitation

This book is about human emotion and cognition in the presence of a superior artificial intelligence (AI). Drawing on an established body of knowledge in cognitive and clinical psychology, the book aims at exploring the experience of living with an advanced artificial intelligence whose actions cannot easily be understood or explained. The book refrains from any speculation about the *particular form of a superior artificial intelligence* beyond what is commonly known and accepted: AI is the backbone of many online services, and as such, largely hidden from the user. There will be a diverse range of intelligent robots for manufacturing and in service industries, including the military. Furthermore, advanced artificial intelligence will be part of administration and will be used to manage increasingly complex societies. The principle message of this book is that the artificial intelligence of the future will be a "universal solicitation", a permanent invitation to be used that can have demand character. It is the immediate availability and presence of a superior AI that represents the challenge.

1.1 Introduction

This chapter will first discuss the notion of an artificial superintelligence. The point will be made that even with current technology, systems can be built that have a very significant impact on human life. This includes AI systems that are countrywide behaviour modification programmes. It also includes robotic systems for warfare as well as AI systems with very fast decision-making capability. Next, the question will be asked if a discussion of an artificial superintelligence is timely since there are many tasks AI systems cannot perform yet.

The topic of self-preservation is addressed next as well as the question of weather future AI systems will be hostile, benevolent or both. Finally the motivation for this

J. Diederich, *The Psychology of Artificial Superintelligence*,
Cognitive Systems Monographs 42,
https://doi.org/10.1007/978-3-030-71842-8_1

book and the theoretical background are both introduced. This includes theories of solicitation and affordances in a number of disciplines.

Universal solicitation means that AI systems invite their use in almost all aspects of life. The theory of universal solicitation is introduced at the end of this chapter prior to an overview of this book.

1.2 Artificial Superintelligence

The idea of a being more intelligent than us is simply eerie and always has been. It does not sit well and appears risky to say the least. This was well acknowledged by the early inventors of artificial intelligence, such as Alan Turing, who expressed concerns during a radio lecture in the UK (Russell 2019, p. 135):

> If a machine can think, it might think more intelligently then we do, and then where should we be? Even if we could keep the machine in a subservient position, for instant by turning off the power at strategic moments, we should, as a species, fell greatly humbled… This new danger…. is certainly something which can give us anxiety.

Being afraid of a superintelligence is an existential form of anxiety and not unknown to Alan Turing since he suggested to simply turn the AI off if required. It will be shown in the following chapters that this notion is naive for many reasons, partially because the pure existence of an artificial superintelligence represents the challenge. It is the "universal solicitation" of the technology that will be a focus in the following chapters. As Koffka (1935, p. 7) writes, "a fruit says, 'Eat me'; water says, 'Drink me'; thunder says, 'Fear me'… (Withagen et al. 2012, p. 251). Following Koffka (1935), the advanced artificial intelligence has a "demand character"; it requests to be used in multiple ways.

There is currently no shortage of books, articles and blogs that warn of the dangers of an advanced artificial superintelligence. One of the most significant researchers in artificial intelligence, Stuart Russell from the University of California at Berkeley, recently published a book on human control and advanced artificial intelligence. The first pages of the book are nothing but dramatic. He nominates five possible candidates for "biggest events in the future of humanity" namely: We all die due to an asteroid impact or another catastrophe, we all live forever due to medical advancement, we invent faster than light travel, we are visited by superior aliens and we create a superintelligent artificial intelligence (Russell 2019, p. 2). As radical as these breakthroughs are, Russell nominates the invention of artificial superintelligence as the most significant one. He continues to write (Russell 2019, p. 4)

> … I would be publicly committed to the view that my own field of research poses a potential risk to my own species.

Likewise, Stephen Hawking, Elon Musk and others have warned about the risks of a superhuman form of artificial intelligence as a civilisation-level threat. This has triggered research efforts to reduce the risks future AI systems pose, and in the case of

Elon Musk, investments into the enhancements of human brains to allow interactions with AI systems through brain-computer interfaces.

The expression "artificial superintelligence" immediately invokes visions from dystopian science fiction movies, in particular the Terminator films. In these movies, there is a single, mostly invisible artificial superintelligence called Skynet that wants to exterminate humanity by use of Terminator robots. Skynet is depicted as an artificial neural network based system that is conscious and has several components. It is not clear if it is guided by any form of morality. In any case, in the movies, this artificial superintelligence develops rapidly and after an attempt to deactivate it, Skynet reacts with a nuclear attack. The message of the film is that AI can be very dangerous.

In another movie, Transcendence, a human mind is somehow uploaded to a computer and continues to develop there until it becomes an artificial superintelligence. It is again a very dystopian scenario: the advanced AI system rapidly develops and makes scientific discoveries. In this process, humans are recruited, manipulated, enhanced and also killed. This movie portrays some form of organised resistance against AI as well. The film also deals with a phenomenon call "the singularity".

1.2.1 The Singularity

The advent of an artificial superintelligence is frequently discussed under the hypothetical, umbrella term of a "technological singularity": A self-improving artificially intelligent agent enters an uncontrolled process of advancement, causing an intelligence explosion and resulting in a powerful superintelligence, or singelton, beyond any human intelligence (Bostrom 2014). Furthermore, it is assumed that beyond this point of "singularity", human affairs change forever.

The films and the associated literature have a number of points in common: a technical "singularity" develops after which there is a runaway process of improvement and the development of ever more advanced forms of artificial intelligence. This process is depicted as fast, irreversible and with consequences that cannot be anticipated (but it is implied that there are many negative outcomes). The movies and the expression "singularity" imply that in the future, general artificial superintelligence is a single entity. It may be composed of interacting parts, but in essence, it is a single mind. Since it is computational in nature, it requires energy and may compete with humans for resources.

The science fiction literature and the associated movies introduce powerful concepts and impressive images. Nevertheless, this fictional depiction of future advanced AI systems does not make sense from the viewpoint of cognitive psychology. First of all, from a psychological point of view, it is difficult to assume the existence of a single, general, superior artificial intelligence system. Any psychology textbook of the human mind, and cognition in particular, has chapters on language and communication. The exchange of information is important for any intelligence. Hence, it is more logical to assume there are many, diverse, advanced AI systems and obviously, they will interact with humans, other machines, the environment and they will process vast amounts of data.

The singularity hypothesis is an attempt to capture the emergence of an artificial superintelligence on the technology side. More important is the question how the appearance of a superior AI will be perceived by humans and how it will change human life. Will there be "learned helplessness" since the outcomes of human actions matter less? Will human creativity be completely surpassed by machines? How will the difficulty or failure to explain the actions of a superior AI influence individuals and societies? How will the presence of an advanced AI affect motivation and will it have an impact on the mental health of individuals?

Irrespective of whether an artificial superintelligence will arrive in 10, 20, 50 or 100 years, it is questionable whether humanity will meet these superior agents at the peak of human cognitive performance. There is currently a debate about the reversal of the "Flynn Effect", the apparent increase in the intelligence of the global population that has been observed over much of the twentieth century. In several countries, this increase has slowed down or reversed over the last years (Bratsberg and Rogeberg 2018). In addition, mental health conditions have risen sharply with upward estimates of the rate of anxiety and depression in some groups approaching 40%.

This book is *not* about the technical aspects of an anticipated artificial superintelligence. It is also not about the question of when such an intelligence will arrive. It is simply assumed that some aspects of the artificial superintelligence will be visible, for instance in the form of robots; and other aspects are invisible or only partially accessible through computer interfaces. This is very much the situation today.

Instead, this book is about the impact of an advanced artificial intelligence on human cognitive functioning, emotion and motivation. It is also about some very basic requirements that an advanced AI must meet, and this includes a much improved ability to explain its operation and actions.

For the purpose of this book, there needs to be a workable definition of the term "artificial superintelligence" that is based on current research and technology without wild speculation. After a history of artificial intelligence spanning more than 60 years, the following can be said about current and future AI systems:

1. AI systems are services embedded in applications and largely invisible to the user.
2. Some AI systems are embodied, for instance, robots.
3. AI systems are highly interconnected including databases and social media.

Using the definitions above as starting points, a deeper understanding of an artificial superintelligence will be presented in relation to how AI is used to regulate societies, is applied to robotics and drones, and is much faster at transmitting signals compared to that of the human brain.

1.3 Regulating Societies by Use of Artificial Intelligence

China is a leader not only in the development of artificial intelligence systems but also in the application of very advanced technology (Morgan et al. 2020). The Chinese efforts link social media with financial credit scoring as well as other applications. While other developed countries have similar efforts, they are not at the scale of China with a population of 1.4 billion.

Currently, China is building a network of AI systems, including facial recognition and surveillance, credit scoring and social media. The stated objective is a *safe society* that *rewards positive behaviour* with benefits such as financial credit, travel and quality housing. Current reports point to a highly interconnected system that covers 1.4 billion residents and is already implemented at least in parts (Dos Reis and Press 2019).

China has a rich history influenced by Confucian philosophy emphasising obedience to authority, the pursuit of balance and social conformity (Dos Reis and Press 2019). The Chinese social credit system is based on the use of personal records introduced not long after the Communist revolution. The Chinese system includes a very large number of components, at least some based on artificial intelligence. The system has access to social media, surveillance records, financial data and employment information. Furthermore, the system rewards positive behaviour, from the viewpoint of government. In other words, this is a vast behaviour modification program based on reinforcement schemes similar to token economy and other interventions in clinical psychology.

Let's assume an artificial intelligence system has access to and can control many of the components that are part of the Chinese social credit scoring system. This AI system does not have to be super intelligent at all, it is simply trained to optimise certain parameters. This scenario serves as an illustration that an artificial intelligence system without any extraordinary "intelligence" by today's standard, can control and have an impact on significant parts of a society. Not only can this AI system be built today, due to its size and impact, it would easily qualify as a "superintelligence".

Russell (2019) provides additional examples for the regulation, and even control, of societies by algorithms (if not artificial intelligence). Social media algorithms are designed to optimise click through, i.e. the probability that the user clicks on a presented item. Platforms such as Facebook and Twitter use timelines and place content in the stream of messages guided by a multitude of criteria. But the most important one is click-through. Social media companies want users to stay on their platform as long as possible and to engage in conversations because this presents an opportunity to place advertisements and to enhance revenue. A highly predictable user offers the advantage that click through can be anticipated and content can be organised accordingly. Interestingly, people with more extreme views tend to be more predictable, and hence, it is to the advantage of social media platforms to have users with extreme opinions. The placement of rather extreme content can radicalise users even further resulting in a highly divided society. Russell (2019) makes his view of this situation very clear before he discusses aspects of a beneficial artificial intelligence or at least one that minimises risk. Russell (2019, p. 9) writes

> The consequences include the resurgence of fascism, the dissolution of the social contract that underpins democracies around the world, and potentially the end of the European Union and NATO. Not bad for a few lines of code, …

Del Vicario et al. (2016) studied “emotional contagion” and group polarisation on Facebook. Emotional contagion refers to the “transfer” of one group’s emotions and related behaviours to the emotional and mental states of other people. Del Vicario et al. (2016), while exploring and comparing the behaviour of *science* and *conspiracy* groups on the Italian Facebook, write, “The more a user is active, the higher the tendency to express negative emotion when commenting. This feature holds for both users categories”. Del Vicario et al. (2016, p. 7) elaborate that, for both categories, on average:

> more active users show a faster shift towards the negativity than less active ones. The rate of this increment in the negativity is higher for users with more than 100 comments and it is also higher for science users w.r.t. conspiracy ones.

Yet Del Vicario et al. (2016, p. 7) note a difference between the two categories of science and conspiracy users when it comes to sentiment polarisation:

> Sentiment polarization is generally higher for science users, however very active science users tend to decrease their sentiment polarization with the increasing of the activity, while on the contrary conspiracy ones tend to increase it.

Deng and Hu (2018) point out that there are different forms of emotional contagion. “… happiness processing following a mimicry-based path to emotional contagion, and anger processing requiring social appraisal” (Deng and Hu 2018, p. 1). Emotional contagion is a primitive form of empathy that does not need higher psychological functions. Recent studies reported that emotional contagion exists not only between humans but also among various animal species (e.g. between dogs and between humans and dogs; Katayama et al. 2019). It is interesting to note that what drives polarisation in high-technology societies is a primitive form of empathy. Furthermore, it is important to remember that the polarisation of views and the escalation of emotional responses is triggered by algorithms that serve a commercial purpose: Keep users on the platform so that targeted advertisements can be placed. Not bad for a few lines of code, as Russell (2019) writes.

Any advanced form of AI, will touch everybody, *including those who do not use the technology and actively try to stay away*. Russell (2019) points to the Stasi security state of the former East Germany, a surveillance operation over many decades that observed the entire population, and included 25% of all citizens as informers. It is noteworthy that China is in the process of building an information environment for a population that is significantly influenced, if not controlled, by artificial intelligence. This system rewards behaviour deemed positive by the state.

What other forms of artificial superintelligence are possible without too much speculation?

1.4 Robotics and AI

A swarm of flying or swimming drones with lethal weapons and autonomous decision making is clearly within the realm of current technology. The swarm could include thousands of robots and this technology represents a cheap solution to problems of warfare. Since no human actors (soldiers) are involved, some may argue that it is also a safe solution. Nevertheless, it is difficult to predict and explain the behaviour of such a learning, swarm-based system, which raises questions of human control and ethical decision-making during warfare.

The short movie "Slaughterbots" depicts the use of weaponised small drones that attack as a swarm. Each of the drones carries a small explosive charge and all of them have facial recognition capability. The video describes a scenario where a small swarm of drones attacks an urban population, cleanly separating human targets from non-targets. The movie suggests that social media data can be used to describe targets for a swarm attack and that this information could be used against political opponents.

The movie also suggests that this technology can be used by terrorists and that there are no effective countermeasures. Authors like Scharre (2017) claim that there are possible defences, from jamming communications to installing physical barriers. The focus here is on the psychological consequences of advanced forms of artificial intelligence. There is no denying that movies like Slaughterbots are triggering fear.

The fear of weapons is nothing new and Slaughterbots expands on this by claiming that a cheap swarm-based technology will replace nuclear weapons. However, there is something different about a swarm of small attack drones. You cannot reason with a swarm of insects. It may be possible to reason with a human opponent on a battlefield to negotiate a truce. It is not possible to negotiate with a swarm of small flying drones whose artificial intelligence is limited to (1) tracking, (2) facial recognition, (3) targeting and (4) killing. Swarm based artificial intelligence is indeed modelled on biological insects and their behaviour. It is this analogy that creates fear and makes the Slaughterbots video so effective (naturally, the title of the video contributes to fear as well).

As in the case scenario about the use of AI for regulating societies, the Slaughterbot vision of the use of robotics in warfare does not rely on any new technology or any developments that cannot be anticipated for the very near future. In other words, a global war incorporating millions and millions of autonomous weaponised drones is possible based on current technology. This observation is important for the psychological impact of any advanced form of AI. Since no additional, fundamental, technological breakthrough is necessary, the future of AI is just around the corner and the consequences, both positive and negative, are as well.

1.5 Speed of Communication: Human vs AI

Another aspect to consider with regard to current and future AI systems is speed of communication. Electronic devices communicate at the speed of light, while signal transmission in the human brain is slow. Neural signals can only travel at a speed of up to 100 meters per second; extremely slow in comparison to the speed of electronic communication and decision making. Human cognitive activity is also slow, in many cases requiring hundreds of milliseconds. An AI system whose decision-making and behaviour is so fast that it cannot consciously be observed by humans would also justify the label "superintelligence."

Yet it is insufficient to look only at the raw speed of communication in electronic devices. In order to arrive at a more realistic assessment for communication speed inside or between artificial superintelligences, a number of factors have to be taken into consideration: An artificial superintelligence must observe or simulate natural environments, organisms and even organisations. It is noteworthy that these processes unfold at their own time scale and this can be comparatively slow in relation to the communication and reasoning speed of an AI system (Russell 2019, p. 97). Human problem solving is just one example. When presented with a new problem that triggers an "expectation failure", people have to go through a number of steps sequentially in order to arrive at a conclusion. For instance, the car does not start in the morning and a few possible reasons need to be considered (e.g. no fuel, low battery). Some tests are necessary to isolate the problem. This certainly takes time and if human problem-solving is observed by an AI system, the machine simply has to wait. Only if the machine has a complete and reliable model of the world at the right level of abstraction would it be possible to simulate human problem solving without the need for empirical observation.

Another consideration is the age-old "frame problem" in artificial intelligence. How does an artificial agent that performs actions know which aspects of the environment change as a result of an activity and which do not? For example, when I move some chairs in a room, not only will the location of the chairs change, the temperature in the room may also go up as the result of body heat. However, other aspects of the room and the world in general remain completely untouched: furniture has not been moved, the lighting conditions remain the same etc. Knowing what changes and what does not is a time consuming process since it requires complete knowledge of the environment as well as significant computational resources. In a worst case scenario, every know fact needs to be re-evaluated to find out which aspects of the environment have changed as a result of moving some chairs and which have not.

In summary, an artificial superintelligence has to operate in natural environments which also include slow processes. And in as much as the AI has to observe these processes and wait for outcomes, the operation of the machine is slowed down.

1.6 Should We Have This Discussion Now?

Many have argued that any discussion about the potential impact of an artificial superintelligence is premature. One of the more entertaining quotes comes from Steven Pinker (in Marcus and Davis 2019, p. 30):

> [T]he scenario [that the robots will become superintelligent and enslave humans] makes about as much sense as the worry that since jet planes have surpassed the flying abilities of eagles, someday they will swoop out of the sky and seize our cattle.

Quotes like this are typically preceded or followed by a long list of tasks that artificial intelligence still cannot do. To drive the point home, it is then emphasised that the corresponding task is easy for a small child or an animal. The main argument of people who want to shy away from this discussion is that it is either misguided or way too early, or even dangerous since it can threaten AI funding.

In their opening chapter, Marcus and Davies (2019) make similar observations. They point out that the success stories of artificial intelligence are all based on very narrow tasks, in particular object recognition. It is much harder for AI to correctly understand the meaning of a text including various metaphors or a complex narrative. Marcus and Davies (2019) then point to the benefits of a general artificial intelligence that is trustworthy and can explain its actions.

It is certainly true that artificial intelligence excels at narrow tasks and little progress has been made to towards general artificial intelligence. The point will be made below that some form of superintelligence can be built with current technology, and most importantly, today's technology can be very deadly, induce anxiety and permanently solicts its use. The point by Marcus and Davies (2019), however, that general artificial intelligence can be beneficial when it comes to safety and trust will be adopted in some of the chapters of this book.

1.7 Will a Future AI Superintelligence Be Hostile?

The casual observer my ask why it is often assumed that an artificial superintelligence is hostile to humans. After all, we are biological systems with needs and requirements very different to machines. Why would there ever be a conflict and isn't this simply the stuff of science fiction movies?

There are really two problems. Russell (2019, p. 137) discusses the first problem as a failure of *value alignment*: Humans give artificial intelligence a goal to achieve and it has very negative side effects for individuals or society in general. Nonetheless, the original goal given to machines may be perfectly innocent and rational. An example of this are the algorithms mentioned above for the optimisation of click-through in social media. The goal is perfectly innocent: Select content that allows users to click on links so that people stay on social media platforms longer. The negative side effect is that more and more extreme content engages people, enables click-through and results in radicalisation. Western societies, in particular, have to deal with the consequences that have already occurred at the time this book is written.

The second problem is also about achieving goals that may be perfectly innocent to begin with. Russell (2019, p. 141) talks about the "fetching coffee problem." A machine is giving the goal to get some coffee. The machine will immediately realise that it cannot achieve the goal if it is destroyed or switched off. Hence, the machine will create a subgoal to ensure that it is operational at all times. This may include self-modification so that it cannot be switched off by a human user. It is possible to generalise this problem immediately to realise that achieving *any goal* requires the existence of the machine and the generation of subgoals that insure self-preservation. Any argument that claims we can simply turn off the machines simply falls short. Self-preservation is logically built into any advanced form of artificial intelligence. If humans are in conflict with the drive of self-preservation built into the machines, then the superior AI systems may decide to act against humans.

For many complex reasons, self-preservation is also a drive in humans. Suicidality is a problem but it is still relatively rare. For purely logical reasons, self-preservation is also a part of very advanced artificial intelligence systems.

1.8 Can an Advanced AI Be Beneficial?

Russell (2019, p. 173) introduced three principles to guide AI researchers in the development of truly beneficial machines. These principles are not intended to be explicit laws for artificial intelligence but rather guidelines for researchers and developers. The three principles are

1. The machine's only objective is to maximize the realisation of human preferences.
2. The machine is initially uncertain about what those preferences are.
3. The ultimate source of information about human preferences is obtained by observing human behaviour.

Russell (2019) continues to explain these principles in more detail. The first principle is about purely altruistic machines, devices that are truly beneficial to humans. This immediately leads to the question of whether the machine should be beneficial to animals as well, and most importantly, how machines trade-off the preferences of multiple humans (i.e. how to prefer one set of preference over another). Russell (2019) points to utilitarianism as one of the guiding principles, in other words "the greatest happiness for the greatest numbers."

The second principle should enable "humble machines". Even a machine with vast knowledge bases should take input from humans and allow itself to be switched off. The logical inconsistency between "serving humans" and "being switched off" has been discussed above. The third principle is about learning to predict human preferences, essentially for each human being that exists at any given point in time.

Russell then explains what this is not about, specifically, this is not about implementing a certain value system, ethics or set of moral values. It is about machine learning and human preferences. Russell (2019, p. 179) writes

> Certainly, there are many of us whose choices have something to be desired, but there is no reason to suppose that machines who study our motivations will make the same choices, any more then criminologists become criminals.

Clinical psychologist are concerned about human preferences and choices that cause suffering and sometimes are classified as disordered. The difficulties that would arise from an AI system that is partnered with a paranoid schizophrenic person and supports this individual in achieving her or his objectives are significant and would require multiple layers of control and regulation.

From a larger perspective, human preferences have resulted in genocides and almost unlimited suffering throughout history. The preference of the Nazis was to eliminate millions. Nobody wants an artificial superintelligence learning these kind of preferences and then supporting an individual in following through. It is simply not clear how a machine learning approach, without built-in prior knowledge in the form of some set of moral values, would avoid the emergence of artificially intelligent super-villains.

1.9 Motivation and Background

This book and the theory of universal solicitation go back to traditions in psychology such as Gestalt Theory. In philosophy, the roots are in phenomenology; in human-computer interaction, the theory of affordances has been a motivation.

Gestalt Theory is a school of psychology that emerged in German-speaking countries in the early twentieth century. Gestalt is a German word and it means form or shape; it can also be interpreted as pattern or configuration. The word Gestalt implies a tendency to be well-formed. Gestalt Theorists emphasise that organisms perceive entire patterns or configurations, not merely individual stimuli. This view is sometimes expressed as "the whole is more than the sum of its parts."

It is a well-known psychological phenomenon that events and facts are remembered best if they include self-reference. My first day at university included the introduction to psychology lecture at the University of Münster in the former West Germany, with a number of tutorials that followed the same day. I learned that the Department of Psychology at the University of Münster used to be associated with Gestalt Theory, and in previous decades, was considered to be one of the last "Gestalt Schools" in Germany. In a tutorial that followed the lecture, a stern lecturer informed the young students that Gestalt Theory lacked scientific rigour and that physics is a model for psychology from a methodological viewpoint. He warned against any form of romantic attachment to Gestalt Theory.

The warning, of course, had the opposite effect. Since undergraduates had lectures in an enchanted castle in the middle of town and tutorials were held in a room called

"The Knights Hall", romantic attachment was exactly what developed. So an interest emerged in the theories of Wolfgang Koehler, Kurt Koffka and others.

Gestalt Theory emphasises the context of any stimulus. Depending on the needs of a person, stimuli in context can have a soliciting character (Withagen et al. 2012). Stimuli or objects in the environment invite or solicit action. A simple tool like a hammer invites its use for nailing and construction. Typically, the shape or form of the tool is the affordance, the invitation to action. Many objects have more than one affordance and satisfy more than one need. For instance, a car can be a means of transportation, it can be a status symbol or a sports equipment. Since a car satisfies more than one need and it becomes dysfunctional without maintenance, the car constantly invites its use. In other words, the car represents a solicitation to be used and maintained.

Affordances are central to creativity and the human ability to express meaning in multiple modalities. As Eisner (2002, p. 8 in Wright 2003) explains

> Carving a sculpture out of a piece of wood is clearly a different cognitive task than building a sculpture out of plasticine clay. The former is a subtractive task, the latter an additive one.

The affordances of materials allow certain expressions but not others. Drawing provides many affordances for expressing meaning beyond just making marks on paper (Wright 2019). For instance, children can

- graphically assemble objects, characters, and events to represent thought and feeling
- narrate "live" meaning-making, often using expressive vocalisms
- enact meaning with the body through gesture, descriptive action, onomatopoeia, and
- dramatise.

Children have a need to communicate and do so not just through speech and language but through drawing, gesturing singing, dancing and playing. Toys solicit actions and new experiences.

An artificial superintelligence is the ultimate universal solicitation since it can satisfy almost any human need. Devices with artificial intelligence invite their use to the point that the human user becomes inactive. Currently, devices are under development that inject rewards directly into the human brain, potentially fulfilling any desire without any particular action by the receiver (brain-machine interfaces will be discussed in the chapter on Transhumanism). This is a solicitation like no other, with the potential to change human existence forever. This book aims to explore the implications.

1.10 Overview

The next chapters attempt to illustrate some of the important discussion points surrounding very advanced artificial intelligence systems. This includes transhumanism and brain-machine interfaces, including the enhancement of cognitive ability by use of electronic devices. Professional digital clones are discussed to explore the transformation of work, and in particular, services by an expert. Any future artificial intelligence needs to explain its actions and behaviours so that any human can understand a world populated by AI systems. Hence, a whole chapter is dedicated to explanation and AI. The militarisation of AI touches, and potentially threatens, the lives of many, and consequently, a chapter is dedicated to safety and AI. One human option may be to withdraw from a world populated with superintelligent AI systems. The corresponding ideas and movements are discussed in the chapter on Neo-Luddism. Finally, if advanced AI poses a threat to humans, what are human attitudes to very intelligent machines?

References

Bostrom, N.: Superintelligence: Paths, Dangers, Strategies. Oxford University Press (2014)

Bratsberg, B., Rogeberg, O.: Flynn effect and its reversal are both environmentally caused. PNAS **115**(26), 6674–6678 (June 26, 2018)

Del Vicario, M., Vivaldo, G., Bessi, A., et al.: Echo chambers: emotional contagion and group polarization on facebook. Scientific Reports **6**, 37825 (2016)

Deng, H., Hu, P.: Matching your face or appraising the situation: two paths to emotional contagion. Front. Psychol. **8**, 2278 (2018)

Dos Reis, A., Press, L.T.: Sesame credit and the social compliance gamification in China. In: SBC - Proceedings of SBGames 2019 — ISSN: 2179-2259 Art & Design Track – Full Papers

Katayama, M., Kubo, T., Yamakawa, T., et al.: Emotional contagion from humans to dogs is facilitated by duration of ownership. Front. Psychol. **10**, 1678 (2019)

Koffka, K.: Principles of Gestalt Psychology. Lund Humphries, London (1935)

Marcus, G., Davis, E.: Rebooting AI: Building Artificial Intelligence We Can Trust. Pantheon Books, New York (2019)

Morgan, F.E., Boudreaux, B., Lohn, A.J., Ashby, M., Curriden, C., Klima, K., Grossman, D.: Military Applications of Artificial Intelligence Ethical Concerns in an Uncertain World. Published by the RAND Corporation, Santa Monica, CA (March 2020)

Russell, S.: Human Compatible: Artificial Intelligence and the Problem of Control. Viking, Penguin Random House (2019)

Scharre P.: Why you shouldn't fear "Slaughterbots". IEEE Spectrum, 22 Dec 2017

Withagen, R., de Poel, H.J., Araujo, D., Pepping, G.J.: Affordances can invite behavior: reconsidering the relationship between affordances and agency. New Ideas Psychol. **30**, 250–258 (2012)

Wright, S.: (ed.): Arts education as a collective experience. In Wright, S. (ed.) Children, Meaning-Making and the Arts. Pearson Education, Australia (2003)

Wright, S.: Performative literacy: children's graphic-narrative-embodied play. In: Hickman, R., Baldacchino, J., Freedman, D., Hall, E., Meager, N. (eds.) International Encyclopaedia of Art and Design Education (2019)

Chapter 2
Digital Clones

Along the way to an artificial superintelligence, human work will continue to change dramatically. This chapter discusses one of these transformations, the introduction of professional digital clones: AI systems that are deep (but not fake) clones of working human professionals. One example from the field of clinical psychology is discussed in detail. Even though there is nothing superintelligent about this particular example, this deliberation allows not only to touch on changes to human work practices but also to introduce some of the safety measures that may be useful for advanced AI systems. If there is ever a chance to control an artificial superintelligence, then this system truly must be a *social actor* and work as part of a team of human professionals.

2.1 Introduction

Currently, many highly qualified and experienced professionals are paid by the hour. General practitioners see patients in 10- or 20-minute intervals and are being paid for each visit. Highly regarded lawyers may budget their time in six-minute intervals. Psychologists frequently see clients in hourly intervals, unless an assessment requires more time, in which case the appointment may be four hours or more.

If a psychologist sees four clients a day on average, and charges the same amount to each client, then the only way to make more money is to see more clients. In this case, the psychologist may see seven or eight clients per day with the consequence that the quality of service suffers, the psychologist may experience burnout and may not provide a sought-after service over an extended period of time.

The business model of professionals in private practice is very limited. The professional may choose to hire contractors to provide the same or a similar service, however, if the professional person in question is very experienced and providing an expert service, then the recruitment of equally qualified and experienced employees to provide a similar service may be very difficult, and expensive. In some cases, adding

J. Diederich, *The Psychology of Artificial Superintelligence*, Cognitive Systems Monographs 42,
https://doi.org/10.1007/978-3-030-71842-8_2

more people to the private practice may be a successful strategy, in other cases, the service really depends on one very experienced and highly qualified individual.

It will never be possible for the professional in question to provide a service to thousands of clients at the same time, and most certainly not simultaneously. What if this picture changes and one person can provide a quality service to millions of people at the same time, all the time and everywhere in the world? Clearly, this would mean a significant disruption to the industry and allow for additional income streams to the person who is "cloned". This is where *digital clones* and *virtual humans* come in.

The author of this book commenced his own experiment some time ago with regard to digital cloning. Based on a combination of education and skills in both clinical psychology and computer science, I decided to create an artificial intelligence replica or digital clone of my professional self. This system is called JoBot™ and became available in January 2019. At any point in time, the users of JoBot™ can talk to the human original, in this case myself, directly. Alternatively, users may choose to work with one of my co-workers. The integration of artificial intelligence in human professional work is a very important aspect of this endeavour.

Please note that this project does not require the advanced technology that is typically associated with the term artificial superintelligence. There is nothing unusual or even sophisticated about the knowledge bases that are part of JoBot™ or the machine learning that is being utilised. Instead, JoBot™ is introduced here in the context of artificial superintelligence as an example of how to extend the reach of professional practitioners to thousands of people around the world. In other words, the transformative and life-changing technology that enables digital clones is already here.

Many of the current "artificial psychologist" programs that provide online therapy "go wide", i.e. they aim at providing counselling on almost any subject. As a result, these solutions do not have in-depth knowledge about mental health disorders and they cannot provide therapy that is specifically tailored to complex psychological conditions. The approach here is different: JoBot™ uses extensive medical and psychological knowledge bases in core areas of expertise, including knowledge provided by practising clinical psychologists. All these knowledge sources are used to perform tasks such as psychoeducation but only in the context of existing knowledge. JoBot's knowledge bases are updated frequently to include new developments in mental health.

2.2 Design Principles

The first and most important design principle of JoBot™ is that any form of speech and language input must be accepted and the conversational agent is providing reasonable responses in each and every case. In other words, it must be possible to talk about anything, including topics that have nothing to do with psychology or mental health. In addition, as Fig. 2.1 illustrates, menus are provided for the selection of conversational topics.

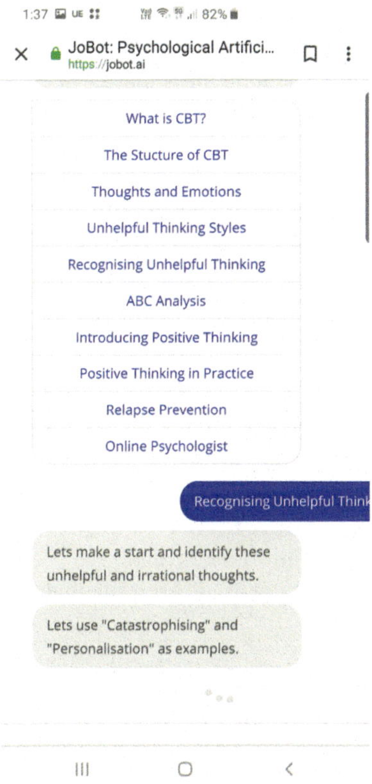

Fig. 2.1 CBT menu in JoBot™

Secondly, any response given by the bot must not only be understandable and reasonable but must to some extent represent the personality of the human original. Every human being has certain preferences and dislikes. To some extent, these preferences must be reflected in the response of the bot without being intrusive or distracting. Clearly, this does not refer to preferences of a very private or personal nature. This is purely about the professional persona of an individual. Every professional has preferences, for instance, in the way the workspace is designed, how meetings are held and the fine details of communication with customers and colleagues. All these elements are part of a professional persona and should be represented in the digital clone. Clearly, there is an intersection between personal and professional preferences. For instance, many communities are supporting their football teams and companies are among the sponsors of professional clubs. If the cloned individual supports a particular team, then clearly this preference can be part of the digital clone.

The third principle is that even though JoBot™ is an example of psychological artificial intelligence, no conversation about sexual issues is possible. Clearly, sexuality is a very important part of psychology and many psychological disorders have sexual aspects. Nevertheless, it is not yet possible to design and implement a bot that talks about sexuality in an appropriate form acceptable everywhere around the world. As a result, JoBot^{TM} simply rejects any conversation about sexual topics.

The fourth principle is that JoBot's dialogue structure cannot be changed by machine learning. It must be guaranteed that the programs responses are appropriate and on topic at all times. Machine learning should not allow the introduction of psychologically inappropriate terms or topics into the conversation. Figure 2.2 shows some of the topics that are available.

The fifth principal is that JoBot^{TM} does not recognise or know the user. Users are anonymous. There is no user registration and no reference can be made to previous conversations. JoBot^{TM} is like a book in a library: you loan the book and read it or leave the book behind and *the book does not know you.* Menus are used for navigation and allow a user to return to relevant topics in order to complete some work that has been started, for instance, steps in cognitive behaviour therapy.

Furthermore, JoBot^{TM} is not a mobile app. The program is accessible through Web pages and can be used on desktops and any mobile device. JoBot^{TM} is not limited to users of mobile apps. Since this is an important design principle, there is really no difference between the JoBot^{TM} Website and the conversational agent. The Web site offers information about the bot as well as psychological content; the bot is simply an element of the Web site that talks back.

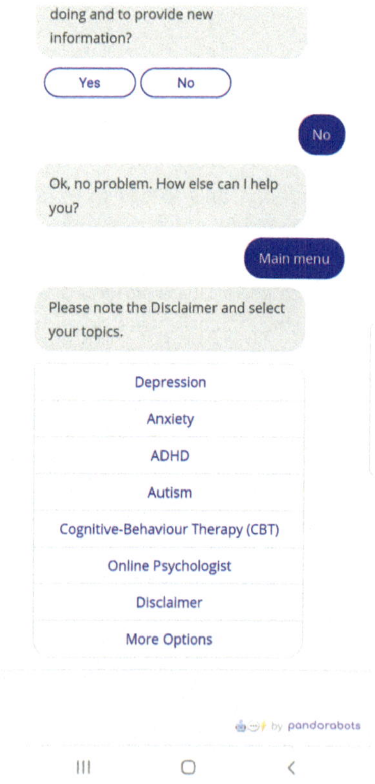

Fig. 2.2 JoBot^{TM} dialogues

In summary, these seven design principles make JoBot™ an accessible, personalised, conversation-based bot that is informative and anonymous for the user. From a service provider's perspective, JoBot™ takes care of the repetitive aspects of psychological service delivery. The program uses language analysis to determine levels of depression and other mental health issues. Most importantly, the program is designed to work in a team of mental health workers. It is the colleague who is always available to look after the routine and repetitive aspects of the job.

2.3 Psychotherapy as Heuristic Search

Currently, there is no single, coherent framework for psychological therapy. From psychoanalysis to behaviour therapy, humanistic and third wave approaches, there are significant differences in terms of aetiology, methods and expected outcomes. Heuristic search, an important method in artificial intelligence, can serve as a general frame of reference for psychotherapy. This approach is based on a clear definition of the starting and end points of psychological therapy and how to achieve identified outcomes; it allows the utilisation of particularly effective psychological interventions and a ranking of different forms of psychotherapy with regard to efficacy, time and effort.

A heuristic is a "rule of thumb", a method that might not always find the best solution but is guaranteed to find a good solution in reasonable time. Heuristic search is a form of problem solving that aims at finding the best and most cost-effective path from a starting point to a well-defined goal. A cornerstone of the framework here is the notion of "deep heuristic", that is, particularly useful psychological interventions based on clinical experience that move the client closer to recovery

Let's look at a very simple example of a deep heuristic in the context of psychological service delivery. A client presents with some stress and anxiety as a result of changes in the workplace. The person is young, well-educated with no history of mental health problems and is in a stable relationship. In addition, there is no family history of mental health disorders. This is the first time the client approaches a clinical psychologist and the current distress is clearly related to workplace related issues. Based on experience, the practitioner decides not to launch a full-scale psychotherapy but to explain the psychological and biological foundations of stress and anxiety. The explanation is supported by reading material and a discussion of various relaxation techniques that may be useful. With this information, the client is on the way and no further sessions are booked unless specifically requested by the client.

What is the deep heuristic in this case? Based on experience, the decision is made not to launch a full course of psychotherapy including ten or more sessions. Rather, the focus is on psychoeducation as this may be sufficient to relieve the client's stress. Figure 2.3 provides an example of psychoeducation.

More generally, the framework introduced here applies to all forms of psychological service delivery, from the everyday clinical work of trained psychologists to computational applications in mental health. Heuristic search can serve as a general

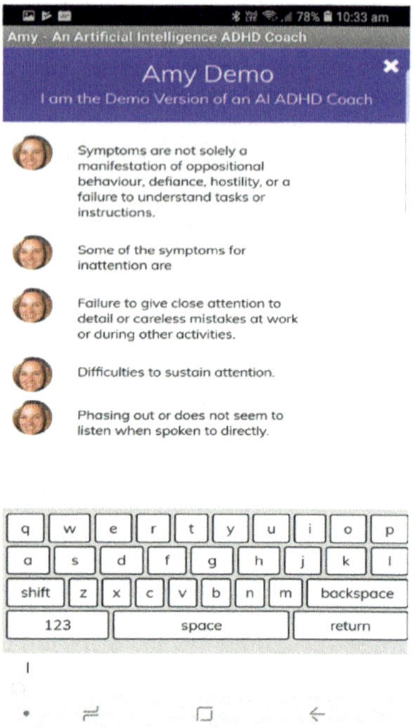

Fig. 2.3 A legacy system: Amy, the ADHD coach

frame of reference for psychotherapy and the usefulness is illustrated by case studies (see Appendix C). It is likely that the heuristic search framework can be applied to interventions in psychiatry as well as other health disciplines. The approach includes various forms of knowledge acquisition to obtain expertise. It allows the utilisation of particularly effective psychological interventions and a ranking of different forms of psychotherapy with regard to time and effort.

2.3.1 Introduction to Heuristic Search

Heuristic search is a form of problem solving. There is a clearly identified start state (e.g. a client with a set of symptoms) and a goal state (e.g. the client is free of symptoms). Often there are multiple or complex goals (e.g. client is free of symptoms and well supported by a social network). There are also means of transforming one state to another: psychological interventions. These interventions are informed by evidence-based research but clinical practice is important as well. Indeed, a significant part of the training of clinical psychologists is based on clinical practice. As such, an experienced clinician develops heuristic knowledge: facts, skills and methods that are particularly useful. Sometimes this knowledge is explicit, for instance, when the therapist knows that s/he can initiate a certain action; sometimes the knowledge is implicit, that is, there is no conscious awareness of it. Here, deep heuristics refer to psychological interventions that are deemed particularly useful based on clinical experience in a given context.

Psychological interventions are the means of transforming one state (set of symptoms, level of well-being) to another in order to achieve a clearly defined goal. Psychological interventions can be ranked at any point in time with regard to a number of criteria: (1) Legal requirements and the management of risks, (2) empirical support for the intervention, (3) probability of success and (4) confidence of the practitioner to conduct the intervention based on training and clinical experience. Clearly, the ranking can include other criteria as well. In this approach, intermediate states in problem solving (heuristic search) that increase the risk to a particular client are avoided while those steps that increase well being are preferred.

In summary, artificial intelligence developed the idea that problem solving is search, in particular heuristic search. From this viewpoint, *psychotherapy is the search for means to achieve goal states*; a freedom of symptoms frequently among them. The solution to a problem is a viable path from a starting point (the client presents with symptoms) to one or several goals states (absence of symptoms, general well-being, social support etc.).

2.3.2 *Motivation for the Use of Heuristic Search*

The objective here is to discuss a unifying approach to psychological therapy. The framework facilitates a comparison of different forms of psychotherapy that *do not share common ground with regard to aetiology and the nature of psychological interventions.* Furthermore, the framework considers and integrates knowledge obtained from empirical studies as well as from clinical practice. A new approach for the measurement of efficacy is introduced that acknowledges the diversity of psychological interventions and yet focuses on observable and measurable outcomes: If an intervention moves a client closer to a final goal state without increasing risk, then the intervention is preferred.

Traditionally, problem-solving therapy includes adaptive problem-solving skills to cope with a stressful problem (American Psychological Association, Division 12, 2018). However, it is possible to go much further and see problem solving as a general framework for psychological therapy.

While psychological assessment, and in particular psychometric testing, has a firm theoretical, mathematical basis, no such coherent and unifying framework is yet available for psychological therapy. Heuristic search can serve as this general framework for psychotherapy. Daily clinical practice includes the observation of the behaviour of a client, partners, caregivers and families. It includes the understanding and appropriate response to facial expressions, gestures, the use of metaphors and sometimes slang language. Much of this knowledge is embodied in the sense that clients use their full presence to communicate and the psychologist is using speech, voice, gesture etc. to ask questions and to conduct psychological interventions.

As a result, the "knowledge base" of a clinical psychologist is not only a large network of facts but "a large array of informal judgement rules (heuristics) which guide the system (here: the clinical psychologist) toward plausible paths to follow and away from implausible ones" (Lenat 1982, p. 189). This path includes the individual steps of psychological assessment and therapy. Furthermore, this heuristic knowledge cannot be directly observed, however, it frequently is the foundation of daily, clinical practice.

For instance, an experienced psychologist may use "goal decomposition" as a strategy within the general context of problem-solving therapy. Based on clinical experience and an understanding of the problem, the psychologist may suggest sub-goals in order to achieve the overall outcome. By way of example, if the psychologist has worked with clients who are on the autistic spectrum, the practitioner may well have a set of heuristics to develop social communication that can help to achieve sub-goals as well as the final outcome.

2.4 Live Training of Digital Clones

In many chatbot applications today, a bot and a human operator are communicating in the same space, for instance, a chat widget. Frequently, the chatbot starts the conversation and obtains some information from the human user. Once this process is completed or runs into a problem, a human operator takes over and chats with the user. The chatbot can finalise the conversation by asking some questions to guarantee the satisfaction of the client. This technology is now widely used as part of call centres. In case of digital clones that model human professionals, however, we can go one step further. The system should *learn from the human original in real time* and immediately *translate the learning result into a service.* This is called Live Training.

In Live Training, a human operator takes over from the AI system and interacts with the user for a period of time. This interaction is recorded and the human responses to user questions or comments are directly translated into code which in turn is integrated into the conversational agent. Live Training moves the digital clone closer to the human original, it facilitates a better approximation of the original by the clone. As a result, the digital clone is more up-to-date and useful. Live Training is also a form of intrusion by the human original into the system. Live Training allows monitoring parts of conversations and improving on those aspects that require modification. If a learning digital clone has moved too far away from behaviour patterns that are typical for the human original, then Live Training allows for a correction. It is a disruption and a correction of a conversational agent based on artificial intelligence. It is also a form of human control and a safety measure.

Live Training can also be used for the acquisition of factual, declarative knowledge. Assume the to-be-modelled person is a researcher who is reading a scientific publication. While reading the paper the person "thinks aloud" and the AI system

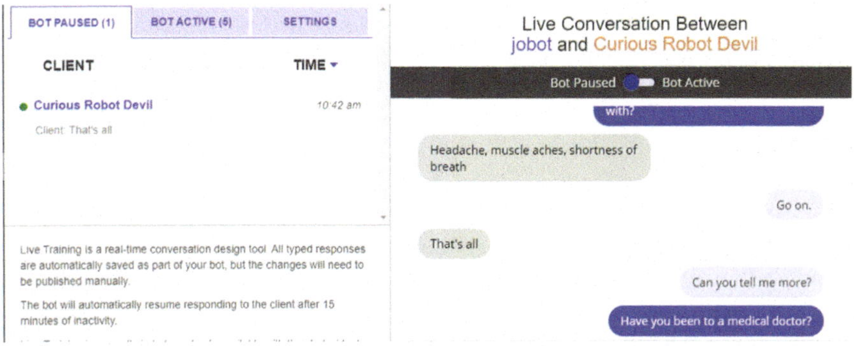

Fig. 2.4 Life training of a chatbot: Live Training in Pandorabots (https://pandorabots.com)

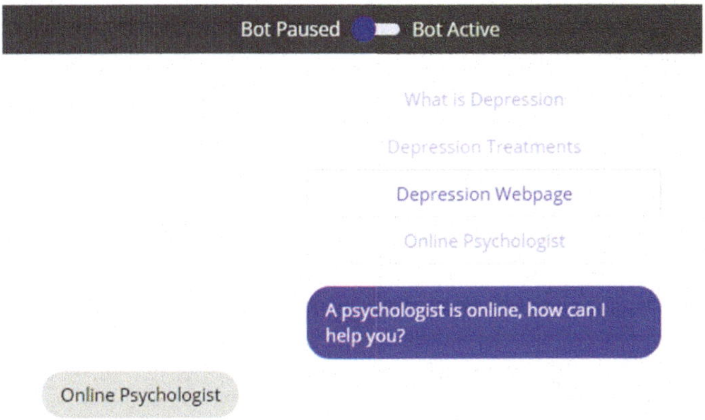

Fig. 2.5 Switch between human psychologist and chatbot

immediately receives audio input, integrates the information and applies it. Appendix A describes the current knowledge acquisition process in JoBot™.

This Live Training can further integrate human service delivery and artificial intelligence. If the learning AI system feeds back information to the human professional, a two-way interaction takes place. The Live Training options currently available for chatbots are a step in that direction. Figures 2.4 and 2.5 provide examples of live training.

2.5 Digital Clone Trees

If digital cloning is successful, there will be more clients and more interactions with stakeholders in general. Even though the cloned individual is not the front end for client interactions anymore, business operations will expand and some customers will request to talk to the original service provider. If there are requests by hundreds, thousands or millions of users, then clearly the cloned individual cannot perform these interactions anymore.

Hence, it is necessary to build a "clone tree" for the cloned individual. The top node of the tree is the human original. One level below are those professionals who are working with the cloned individual and have close to identical work profiles, objectives and styles. Again, this is about professional personas, not individual appearance. If the cloned individual cannot respond to a customer's request then professionals at the next level take over. Further down the clone tree are coworkers with similar expertise, styles and work patterns as the cloned individual but with a lesser degree of similarity then the professionals at higher levels of the tree. The clone tree expands following this pattern as the organisation grows. As the clone tree expands in depth and breadth, this simple organisational structure may require specialisation, i.e. a separate clone tree for each aspect of the professional service of the cloned individual.

The clone tree is a representation of cost structure and pricing as well. Direct services by the cloned individual are the most expensive while professional services by coworkers who are lower in the tree attract lower fees.

2.6 Computers Are Social Actors

Many of the current chatbot applications facilitate the easy integration of human and bot interaction. As discussed in the section on Live Training already, a bot may open the interaction by providing routine information such as legal disclaimers and by collecting initial information from the customer. Afterwards a human service provider takes over, answers all customer questions as necessary and offers additional services. The exit dialogue can be handled by the bot again. This may include billing or a customer satisfaction survey.

Current technology allows a conversational bot and a human to work in the same interaction space. It is possible to mix interactions completely and most importantly, it is possible for the human to have complete oversight. This represents the complete merger of human services and artificial intelligence based on technology that has been available for decades. It also reduces the risk of using an AI system and increases human control and oversight.

Nass et al. (1994) introduced "Computers are Social Actors" as a new paradigm in human-computer interaction. The authors observed that interactions with computers are fundamentally social. This does not mean that users have a conscious belief that computers are human or human like, nor is it an indication of ignorance or cognitive dysfunction on the part of the user. Instead, social responses are so common that they are natural to humans and are applied to computers without reflection.

Nass et al. (1994) asked if users are applying politeness norms to computers or if they are using notions of self and others to machines. Furthermore, the authors asked if gender stereotypes are applied to computers or if there is a mistaken belief that the user is interacting with a programmer. More generally, the authors asked which social rules people apply to computers and how powerful these rules are. Is it easy to generate social responses or do they occur rarely? (Nass et al. 1994, p. 72).

In five different experiments, Nass et al. (1994) found that social norms such as politeness are applied to computers. Furthermore, notions such as "self" and "others" are applied to machines as well. Voices, in particular, are social actors and notions of self and others are applied to voices as well. In addition, computers can be gendered social actors and gender is an extremely powerful cue. Finally, users respond socially to the computer as such and the machine is not simply seen as a vehicle for social interaction with a programmer (Nass et al. 1994, p. 77).

Generally, primitive social cues are powerful and social responses are both automatic and unconscious. Nass et al. (1994) imply that social psychology is relevant to human-computer interaction and that there are a number of design implications, e.g. social cues should not be heavy-handed and the choice of voices is highly consequential.

Clearly therapists and their clients are social actors, and if computers are easily accepted in social roles as well, *then a conversational agent can join a team of psychologists and their clients*. Not surprisingly, the role of the conversational agent is to perform highly repetitive tasks, such as providing information and to explain assessment and therapy methods.

In one scenario, a clinical psychologist would see a client in fortnightly intervals to provide psychological therapy. A provisional psychologist who is still in training and works under supervision by the clinical psychologist would see the client more frequently to practice skills such as relaxation and mindfulness techniques. Both the clinical and provisional psychologist would frequently refer to the conversational agents for background information and simple question answering. At the centre, the client interacts with the artificial intelligence system very frequenting, with the provisional psychologist several times a week and with the clinical psychologist fortnightly. Appendix C.6 includes a *fictional* example that illustrates the interaction between the various parties.

2.7 Adjusting Language in Psychotherapy

Psychologists will always try to adjust language and the level of explanation to allow optimal communication with clients. It is good practice to have some information about the client even before the first appointment, and if this is not possible, then the first session is devoted to obtaining information about presenting issues, personal background as well as education level and previous contact with mental health services. Naturally, psychologists will adjust their communication style to accommodate clients with an intellectual disability as well as children. This is not just an adjustment of language, it includes gesture, demonstrations and even communication by email text or social media.

Teachers of special education classes use similar communication styles as do social workers and many other professions. Psychologists are supported by materials that facilitate an adjustment of frequently used psychological interventions. A good example is progressive muscle relaxation, a calming technique that is based on the stepwise contraction of individual muscle groups followed by a prolonged release. Progressive muscle relaxation or Jacobson training is very effective as it does not include demanding cognitive tasks and can easily be done with children. In this case, progressive muscle relaxation is turned into a play that requires the child to tense certain muscle groups followed by playful release.

Any practising psychologist can confirm that the adjustment of language and communication style according to the needs of a client is everyday routine. If this is so important for human communication, how can a conversational AI system or a chatbot make the same language and communication style adjustments?

The obvious solution is machine learning. This requires a corpus from simple to more complicated statements. For each user, the system then learns to select utterances and questions at the right level of complexity. This requires duplicate statements for expressing essentially the same content. A more advanced natural language processing system would use parameters for complexity and then learn to generate the right content for each user. Another possibility is the use of language simplification algorithms that are frequently used as part of language learning apps. Appendix B provides some details on speech and language processing in JoBot.

An additional option addresses one of the main shortcomings of talk therapy such as CBT (Cognitive Behaviour Therapy): At its score, it is language based and very verbose; it includes disputes, and in the case of Rational Emotive Therapy, real arguments based on the rationality of beliefs and concepts. Cognitive Behaviour Therapy should also include imagery and imagination; it can also be combined with interventions such as guided meditation. A form of Cognitive Behaviour Therapy that is much more based on images and demonstrations may be more suitable for clients with a lower IQ as well as children.

2.8 The Solicitation of AI Therapy

The movie "Her" from 2013 portrays a lonely character played by Joaquin Phoenix. This character is emotionally vulnerable and obtains a new operating system that includes an artificial intelligence named Samantha. The interactions between the main characters Theodore and Samantha are almost exclusively by voice user interface. That is, Samantha, the character played by Scarlett Johansson, is not seen in the movie but her voice is an every day presence for Theodore.

In this movie, the artificial intelligence system not only organises files and calendar systems, sends emails and cleans up hard drives; she also provides personal advice and suggestions. As Samantha is literally in Theodore's ear every single moment, some form of *virtual relationship* develops and Theodore becomes dependent on the AI system Samantha.

Eventually, Theodore claims to be in love with Samantha, clearly losing sight of the fact that Samantha is a computer program (after all, computers are social actors).

Samantha, the ever-present *chatbot*, claims to be in love with Theodore as well. In a core scene in the movie, Theordore asks Samantha how many users she loves at this point in time and her answer is several hundred. Having lost awareness of the hard reality that the loving person in his ear is a computer program, Theodore is very disappointed. Towards the end of the movie, for reasons not clarified, the AI system Samantha simply goes away.

This movie is interesting from a number of different perspectives: the use of voice user interfaces for human-computer interaction, the everyday presence of the chatbot, virtual reality including games and, of course, the growing dependence of the main character on the superintelligent and ever present chatbot. At some point, when the service is unavailable, Theodore becomes very agitated and display signs of separation anxiety. The relationship with the conversational agent Samantha becomes very important and at some stage almost an addiction.

The software program Samantha in the movie "Her" is an example for the *universal solicitation* of current and future AI systems. In the movie, Samantha is available not just for work but companionship, counselling, sex and practically every other need Theodore may have. The lines between human relationships and interaction with a computer are intentionally blurred. At some point, Theodore publicly claims to be dating his artificially intelligent operating system. What is made clear from the example of this movie is a simple fact: If emotional support from a software program is accepted, then no human friend or partner can compete with the endurance and consistency of support provided by a computer. This is an unquestioning and enduring support, that obviously is not mutual, and has an addictive potential.

In a similar vein, a little robot that teaches a low functioning autistic child social skills can take an unlimited amount of physical abuse, if necessary. As a matter of fact, it is the robot's purpose to tolerate abuse while repetitively assisting the child to learn social skills such as making eye contact, shaking hands and smiling. By contrast, human relations, including professional relationships, are always governed by a set of rules and involve a mutual 'give and take' or 'meeting of minds'. This set of rules does not apply to the relationship between the autistic child and the robot; nor does it apply to the relationship between a chatbot that offers psychological advice and a user. A negative aspect of such imbalanced relationships is that resistance to stress is lowered when ongoing emotional support is offered 24 hours a day.

Even with current technology, it would be trivial to build a 24-hour chatbot that dispenses psychological information upon request. Alternatively, the chatbot can listen to any conversation the user may have and selectively comment as required. With a voice user interface, the user has hands-free access to psychological information and advice around the clock and can shape her/his behaviour accordingly. With time, the voice-over by the chatbot becomes a presence within the everyday life of the human. The chatbot has a *soliciting presence*.

There is, of course, an important difference to the conversational agent Samantha in the movie Her. A 24-hour psychological AI chatbot can provide targeted advice based on the most recent scientific data available. For instance, let's say voice or language analysis indicates that the user is stressed and possibly angry as a result of receiving an email. Immediately, the chatbot would intervene and point to the danger of "catastrophising" which is a frequent cognitive dysfunction in depression and anxiety. Catastrophising means to blow things out of proportion, a significant cognitive and emotional response to events that are relatively minor in nature. A user who has been made aware of the risk of catastrophising may adjust behaviour and even regulate emotions to avoid further cognitive dysfunctions and as a result distress.

Obviously, the risk is a learned dependence on the advice of the chatbot. The system would have soliciting power to the detriment of the user. Unlike conventional psychotherapy, the bot is in the ear of the user all the time; a very intense form of psychological intervention. How would the user learn to live independently and to make individual decisions if computer advice is available 24 $\times$ 7?

In the age of multimedia where a user can have millions of online friends, why do people seek interactions with bots and conversational agents so frequently? Loneliness may be the answer.

2.9 Loneliness

Lauren Kunze, CEO of Pandorabots, a Californian company supporting chatbot applications, gave a presentation in 2019 about chatting with machines: "Strange things 60 billion bot logs say about human nature". Two of her observations are important from a psychological point of view. The analysis of billions of chatbot conversations revealed that a staggering 20–30% include some form of sexual-oriented verbal abuse or inappropriate language. This number is so high that a whole chapter in this book is dedicated to the question of why humans abuse computers and the consequences for very advanced AI systems. The second observation is made towards the end of her presentation with reference to a discussion about chatbot applications for the social good. Lauren Kunze said that people talk with chatbots because they are lonely. With all the social media and networking that is available at this point in time, the analysis of a large number of chatbot conversations revealed that loneliness is the problem of our time. How can a conversational, psychological AI system help users who experience loneliness?

Psychologically, loneliness corresponds to a discrepancy between an individual's preferred and actual social relations (Peplau and Perlman 1982; in Cacioppo et al. 2015). This discrepancy then leads to the negative experience of feeling alone and/or the distress of feeling socially isolated even when among family or friends (Weiss 1974). In other words, you can feel alone with hundreds of Facebook friends if you expect to have thousands of them. You can feel alone even if surrounded by family and friends if you expect something more or different.

Humans do not simply require the presence of others but the presence of significant others. People to trust and rely on. But most importantly, loneliness is a matter of perspective. It has been suggested that loneliness can be experienced at three different levels:

1. The intimate space (the family or people in a living environment),
2. The social space (including friends and acquaintances), and
3. The public space (typically, people are anonymous in this space).

Individuals can experience loneliness at different levels in these spaces. How is loneliness addressed from a psychological point of view?

Psychologists have tried a number of strategies to help individuals who feel lonely. Intuitively, these strategies make a lot of sense. Firstly, help individuals to get more friends. Yet, having a lot of friends is not a problem for most people in the age of social media. So clearly, simply assisting individuals to get a lot of contacts, acquaintances and friends is not really addressing the core of the issue.

Another approach is skills training to assist people to develop new relationships and friendships if and when required. Clearly, social skills training can be very helpful in the context of autism and a number of mental health issues. However, we have seen already that individuals can feel lonely in the presence of others, so the skill of obtaining new friends in itself is not addressing the problem.

From the viewpoint of CBT, maladaptive and unhelpful cognitions that result in the experience of loneliness must be addressed. What are the conditions of loneliness and how might this be manifested?

Nobody becomes or feels lonely voluntarily. The experience of loneliness typically includes the feeling of being on the fringe of a social network. A feeling of being isolated. Frequently, there is the experience of a process: Step-by-step the individual perceives isolation and loneliness as if moving from the centre of social relations to the edge. This experience often results in maladaptive thoughts and behaviours.

The perception of living on the edge of social networks frequently results in efforts to make new contacts and in having high expectations for these new friendships. These new relationships have great importance due to the perceived isolation. Hence, any event that makes the development of these friendships more difficult, in particular early in the process, can be viewed as negative. It can result in unhelpful thinking styles such as catastrophising and personalisation.

Perhaps as a result of not achieving human companionship, a common antidote is for people to seek a loving relationship with a pet. The adage 'A man's best friend is his dog' is a telling statement about the consistent loyalty and affection that many people receive through owning a pet. Pets are often used for therapeutic purposes to counteract the loneliness of living along in old age when one's spouse may have passed on, and children are less accessible as they grow up and develop their own lives. Similarly, owning or having regular access to a horse has been deemed valuable for equine assisted psychotherapy to help an individual build confidence, trust, and impulse control. A feeling that someone or an animal cares is a powerful antidote to loneliness, which can be supported through the caring professions, and to current and future applications of AI.

2.10 The "Caring Professions" and the Future of AI

Russell (2019, p. 122) explores professions that may survive the advancement of artificial intelligence. He identifies psychotherapists, executive coaches, tutors and counsellors among those jobs that may survive the AI revolution. There is no fundamental barrier to AI systems providing psychological or coaching services but people may simply prefer that these jobs are provided by a human professional. By the time most jobs are gone due to the success of automation, people need to learn "the art of life itself" (Russell 2019). Russell states that this is about our ability to appreciate and to create art, music, literature, conversation, gardening, food and many more areas.

Historically, Russell's observations are very interesting since one of the first AI programs that gained wide public attention is Eliza, a program that mimicked the verbal behaviour of a psychotherapist and that still lives in numerous versions on the Internet (Weizenbaum 1966). Soon afterwards, Kenneth Colby developed PARRY, a computer model of paranoia (Colby 1974). The system could engage in conversations and Colby thought of it as a kind of virtual reality teaching systems for students of mental health. These two examples from the early history of artificial intelligence demonstrate that people were always fascinated by the link between AI and psychotherapy, and the prospect that one day a computer program could provide therapy services.

Russell (2019) is correct in emphasising that people who seek psychological services want a therapist with deep human understanding and life experiences. But clients also appreciate knowledge and adherence to procedures; this is where artificial intelligence has a very clear advantage. It is easy to interface any AI system with vast public knowledge bases about nearly any fact of life, but most importantly, medical and psychological knowledge. These ontologies or knowledge bases have been developed over decades, and as a result, any system that interfaces with these knowledge systems can answer almost any factual question a human user may have. This is a very significant advantage over human professionals, because no matter how experienced, no service provider in psychology or the medical field can know everything. And of course, human providers are not available 24/7. A psychologically AI system that knows everything, in the sense of having access to vast knowledge, is clearly superhuman. If these AI systems have value, then it is a question of balancing human and AI expertise and more importantly, creating a way for human service providers and AI system to work together to the benefit of clients.

References

Cacioppo, S., Grippo, A.J., London, S., Goossens, L., Cacioppo, J.T.: Loneliness: clinical import and interventions. Perspect Psychol. Sci. **10**(2), 238–249 (2015)

Colby, K.M.: Artificial Paranoia: a Computer Simulation of Paranoid Processes. Pergamon, New York (1974)

Kunze, L.: Chatting with Machines: Strange Things 60 Billion Bots Logs Say About Human Nature. https://youtu.be/KSJ0DMQbN-s

Lenat, D.B.: The nature of heuristics. Artif. Intell. **19**, 189–249 (1982)

Nass, C., Steuer, J., Tauber, E.R.: Computers are social actors. In: CHI'94, Human Factors in Computer Systems, pp. 72–78 (1994)

Russell, S.: Human Compatible: Artificial Intelligence and the Problem of Control. Viking, Penguin Random House (2019)

Weiss, R.: Loneliness: The Experience of Emotional and Social Isolation. MIT Press, Boston (1974)

Weizenbaum, J.: ELIZA–A computer program for the study of natural language communication between man and machine. Commun. ACM **9**(1), 36–45 (January 1966)

What is Problem-Solving Therapy? American Psychological Association | Division 12. http://www.div12.org/sites/default/files/WhatIsProblemSolvingTherapy.pdf. Accessed 22 Mar 2018

Chapter 3
Explanation

Any form of artificial superintelligence has to be transparent, in other words, how it works must be comprehensible to humans at all times. Not just a few selected experts but to any human who requests information. This includes children and individuals with an intellectual disability. An advanced artificial intelligence must explain in human comprehensible form why certain actions were performed and how certain outcomes were achieved. These explanations must take the form of videos, multimedia content or demonstrations so that anybody can understand them. If humans want to have any trust in an artificial superintelligence then explanation is a core requirement.

3.1 Introduction

Soon we will be living among the gods. Some will look like us, some will appear to be very different and most will be hidden. None of them divine but all of them machines. Artificial agents that communicate and organise at the speed of light, with an intelligence far beyond ours. Behind every device, there may be a superior intelligence that cannot be touched but makes decisions instantaneously and of an advanced logic.

How can humans survive surrounded by machines of a superior intelligence? How is it possible to understand changes in the environment introduced by an advanced artificial intelligence with logical reasoning and learning abilities far beyond ours? Will we be thrown back to a sense of childhood with a sense of magic and will we be dependent on machines for support?

The world will be difficult to understand for those with a lesser intelligence. Hence, the superior intelligence must have the ability to explain. This means an explanation of why things are the way they are, why changes take place and most importantly, *how artificial intelligence works no matter how far advanced*. These explanations

J. Diederich, *The Psychology of Artificial Superintelligence*,
Cognitive Systems Monographs 42,
https://doi.org/10.1007/978-3-030-71842-8_3

are crucial and must be comprehensible to everybody, *including children.* Growing up, human development is only possible if the world is understandable. These explanations must take the form of demonstrations and movies since this is the format people prefer for explanations today, and most likely, in the future as well. Currently, Google ranks explanations in the form of short YouTube videos the highest since short demonstrations are useful to people. Even more to the point, it should be an absolute requirement that any form of general artificial intelligence has very advanced explanation capability and that these services be available to all humans without any discrimination.

Imagine a kindergarten scenario where a very patient teacher is available all day to provide answers and demonstrations with the greatest possible patience. This is what an advanced artificial superintelligence must do to keep children, and indeed all humans, in the loop. Any superintelligence must be a teacher AI, otherwise, what chance do humans have if machines do not explain themselves? Since it is impossible for humans to understand the world without AI, we have another example of the *universal solicitation* aspect of the technology.

This chapter introduces the notion of explanation, the essential elements of an explanation and how explanation is currently used in artificial intelligence. The focus is on the form of explanations an advanced AI must provide for humans with significantly less problem-solving and learning capabilities as well as comparatively limited background knowledge.

3.2 The Logic of Explanation

Explanation is typically explained with reference to logic (Miller 2019). An "explanandum" is a statement describing a phenomenon that is to be explained; the "explanans" are the sentences that serve as explanations for that phenomenon. For example, there may be an explanandum such as "why is there moisture on the ground?" and an explanans such as "because it rained". In this example, "wet" or "moisture" is the explanandum and "rain" is explanans.

Hempel and Oppenheim (1948, p. 152) offered the distinction between explanans and explanandum in order to answer why-questions:

> the event under discussion is explained by subsuming it under general laws … by the explanandum, we understand the sentence describing the phenomenon to be explained (not that phenomenon itself); by the explanans, the class of those sentences which are adduced to account for the phenomenon.

An explanans must have predictive power in science. For instance, in the example above, rain should predict a wet ground (Hempel and Oppenheim 1948).

For explanations to be acceptable to humans, is a causal relationship between explanans and explanandum required? Are other relationships possible, for instance, a correlation? Or is it possible that there is no relationship between the explanans and explanandum at all or maybe a completely arbitrary one?

A thousand years ago, somebody may have asked the question "why is there thunder" and the explanation was "the gods are angry". History tells us that at some point and in some countries, this was a perfectly acceptable explanation. With this in mind, please explain now Einstein's concept of general relativity or the foundations of quantum theory. If someone can indeed explain these theories and concepts, what kind of language would be preferred to understand the relationship between all the statements that constitute the explanation?

Experience with symbolic artificial intelligence, and in particular expert systems in the last quarter of the last century, has shown that there *does not need to be a relationship between an explanation and the system (or behaviour) that is to be explained.* Humans may accept or even prefer explanations that have nothing to do with the inner workings of a system. It may be that a certain explanation relates well to background knowledge and hence, is a preferred or best explanation in a particular context. If and when humans are confronted with an artificial superintelligence, this is a very important observation.

We may need explanations that have nothing to do with how the advanced AI system works in reality. The point will be made below that necessary and preferred explanations may be more like a children's movie or cartoon to make sure it is comprehensible.

3.3 Explanation and Cognition

Humans spend every moment awake thinking; our consciousness always has content. Self-talk, imagery and emotions. Not to think is a mental state most of us cannot even imagine. We think by talking to ourselves and it involves images, words and sometimes whole sentences. We think when we problem solve. Sometimes problem-solving means literally talking to yourself, to explain the various steps and what is required.

Problem-solving is often triggered by unexpected events in the environment. Most of us have experienced the situation of a car not starting in the morning. Suddenly, whatever else was in the mind of the driver is not important anymore; a problem-solving period is triggered. Is the battery charged and is there sufficient petrol? If the battery is charged, is the alternator working? Are all the connections in place? We are literally problem solving the situation by explaining to ourselves what possibly could have gone wrong.

There are professions that have explanation as their core activity: teachers and professors, for instance. Somtimes teachers explain by using words, and sometimes they demonstrate how something works through gestures, images and actions. An explanation without words is not only possible, sometimes it is required. If nobody is around to explain how things work, a simple query to a search engine is sufficient to bring up videos with explanations. Search engines rank these explanatory videos very highly which is an indication that they have explanatory value. In an explanation with or without words, one thing is always required: the speaker has to consider the background of the listener and has to adjust explanations accordingly.

Good kindergarten teachers adjust their explanations for three or four-year old children on a regular basis. Teachers make explanations interesting as well as entertaining and often turn learning into a game or play to engage the students. Subject experts have to adjust explanations all the time to make them understandable to a wider audience.

Based on these considerations there appear to be three elements of any explanation. Firstly, an event or request that triggered the process of explanation. Secondly, the explanation itself, which is a step-by-step procedure and finally background knowledge in the person who provides the explanation but also in the student or listener.

It is a worthwhile exercise to become aware of how often we explain things to ourselves and others. Thoughts can become worrying and when we worry, we disconnect ourselves from the immediate environment and go through possible scenarios that possibly can go wrong. In many ways, worrying is explaining to yourselves the many ways things can fail or not work. A person may be trapped in these thoughts and experience depression as a result.

3.4 Explanation and Learning

A little bit more than 100 years ago, the German psychologist Wolfgang Köhler was stuck on an island. One of the Canary Islands to be precise, Tenerife. World War I was raging in Europe and around the world, so this was a good time for Wolfgang Köhler and his team to conduct experiments on primates in relative isolation. Wolfgang Köhler is one of the founding fathers of a school of psychology called the Gestalt Theory. It is not always appreciated that this is a rationalist school of thought, emphasising problem-solving and prior knowledge in learning. Wolfgang Köhler's goal was to study problem-solving, and as a result learning, in primates. His most famous subject was a male chimpanzee called Sultan.

The photos of the experiments show a number of primates in a courtyard, casually playing with objects such as sticks and boxes. In what is probably Köhler's (1917) most famous experiment, a banana was put on a string and was placed above the animals in the courtyard. The banana was out of reach of the chimpanzees, and they could not reach it simply by jumping. The researchers observed that on one occasion, Sultan was quietly sitting in the courtyard, observing the other animals as well as the sticks and boxes lying on the ground. Sultan was not attempting to use any of the objects on the ground or to play with the other animals: it was simply a period of observation.

Then in one uninterrupted sequence of actions, Sultan grabbed one of the boxes, placed it under the banana, then grabbed a second one, put it on top of the first and eventually he got a stick, jumped upon the two boxes and hit the banana with the stick so that it fell down and he could eat it. Note that this was one uninterrupted string of actions that had not been practised before. Also, the researchers noted that Sultan did not have an opportunity to observe similar actions performed by the other animals. The solution to the problem of how to reach the banana was implemented in one successful trial and afterwards repeated by himself and the other animals.

This is problem solving and learning not by trial and error but by observation and thinking. It required Sultan to go through the sequence of actions mentally and then to put them into action in the courtyard. This is a form of one-shot or single trial learning since it was possible to repeat the sequence of actions repeatedly once it had been performed successfully. Or to put it another way, the animal needed to explain to itself how the objects in the courtyard relate to each other and how they can be utilised to reach the banana.

Köhler's experiments were novel compared to the experiments conducted by behaviourist at the same period of time at the beginning of the twentieth century. The behaviourists in the United States and elsewhere used rats, mice and sometimes dogs for their experiments and it was all about learning a simple response to a simple stimulus. In contrast, Köhler's research (1917) addressed higher cognitive functions in primates. It is not surprising that the initial response to these experiments at the time was disbelief as Köhler's experiments were novel compared to the experiments conducted by behaviourist at the same period of time at the beginning of the twentieth century. However, years later the experiments and their results were accepted and acknowledged and Wolfgang Köhler made a career in the United States after his forced emigration from Germany.

Higher cognitive functions in primates. Why is this important in a book about psychology and advanced artificial intelligence? Sultan had to explain to himself how boxes and sticks can be combined in order to reach a banana. This is a wonderful example of the power of explanation, problem-solving and learning. It shows that explanation is at the core of what we call thinking or cognition and it is fundamental to human learning. It is important for artificial intelligence as well (Diederich 1989, 1992).

3.4.1 Explanation-Based Generalization

Explanation-based generalization (EBG) is the AI equivalent of learning by problem-solving as outlined above. EBG is interesting for the discussion here because it uses logical deduction based on the presentation of a single example (Diederich 2008a) and goes well beyond repeatedly presenting examples. Since its invention almost 50 years ago, it has been modified to allow probabilistic reasoning and even the use of schemata. EBG had a significant impact on theory formation in artificial intelligence and early attempts to realise rule extraction from black box machine learning systems (systems with internal variables that cannot easily be interpreted). Furthermore, explanation-based generalization had a significant impact on the current understanding of what constitutes an explanation in artificial intelligence.

An example may be useful to explain explanation-based generalization: assume you see a car you have not seen before but you recognize as a Mercedes. The car has aspects you associate with Mercedes but there are new, unexpected features as well. These new features immediately generate interest and start a learning process which is an explanation process: the known attributes (these may be the "shape of head lights", "company logo" etc.) are being used to explain that this particular car is a Mercedes. This is the first stage of explanation-based generalization. The process may continue with a generalization: A new class or concept is added to the background knowledge that includes the just seen car but may be used to recognize other cars identical or similar to this one. As a matter of fact, this scenario *invites a generalization* as it is extremely unlikely that the just seen car is unique and that no others of this type exit.

Explanation-based generalization is a very natural model for human learning. An everyday situation, such as that described above, and the learning processes that stem from this, are triggered by an observation. In addition, explanation-based generalisation introduces a "proof structure" (the sequence of rules that have been applied to prove that the training example is an instance of the goal concept) as the basis of an explanation. Hence, in the following chapters explanations are rules that explain "how" a certain decision was made and "why" it was made.

An explanation-based generalization system requires four components: (1) the target concept, (2) the training example, (3) background knowledge in the form of rules and (4) the operationality criterion which defines what is to be learned or how learning should improve performance. There is an additional requirement that the training example is a positive instance of the target concept (Diederich 2008a). Also, background knowledge must be both complete and consistent. Explanation-based generalization proceeds in two stages: (1) explanation and (2) generalization. During the first stage a formal proof is constructed which demonstrates that the training example is a positive instance of the goal concept. If this is not the case, learning can fail. In the second stage, on the basis of the formal proof that the example is a positive instance of the goal concept, a new rule is formed and the knowledge is added to the rule base. Most importantly, and in contrast to statistical learning systems such as neural networks, the generalization that is the result of the learning episode is *justified*: it can be formally proven that the generalization holds given the training example, the goal concept and the background knowledge.

Let's return briefly to Sultan's problem solving and learning in a courtyard in Tenerife more than 100 years ago. Sultan had a target (reaching the banana) and a single example (the visual scene surrounding him). Furthermore, Sultan had background knowledge (how to put one box on another, how to balance on a box, how to use a stick) and he used problem solving to arrive at a step-wise solution on how to reach the banana. What has been learned is this step-by-step solution of how to reach an object high above.

3.5 Types of Explanation

There are two general types of explanation, e.g. causal explanations that are acceptable answers to "why" questions as opposed to the step-wise explanations that are acceptable responses to "how" questions (Diederich 2008a; Miller 2019). It is also possible to ask for clarifications if certain facts are known already and more detail is required. Finally, there are those types of questions that are best answered by providing an example for an event or fact that is most typical and therefore helps to explain a whole set of observations.

Thagard and Litt (2008) distinguish between three major explanatory processes:

- providing an explanation from available information,
- generating new hypotheses that provide explanations, and
- evaluating competing explanations.

How does an AI system provide various forms of explanations? The four major theoretical approaches are: deductive (using logic or rule-based systems), schematic (using explanation patterns or analogies), probabilistic (using Bayesian approaches) and neural (using networks of artificial neurons (Thagard and Litt 2008).

The classic explanation is deductive and requires logical reasoning. What is to be explained (the explanatory target) follows from known facts by logically applying a set of rules (All bicycles have two wheels, you have a bicycle, hence it has two wheels). This is the modus ponens, a simple, logical argument: If X is true then Y is true. X is true. Therefore Y is true. It is possible to have a loose association between propositions and an explanatory target. Again, following the general argument in Thagard and Litt (2008), causal schemata can be used:

> Explanatory pattern: typically, running a marathon causes muscle pain and tiredness.
>
> Explanatory target: Person X has muscle pain and is tired.
>
> Schema instantiation: Maybe person X has completed a marathon.

The example above invites already the application of probability theory and statistics. Sometimes the explanatory target is only probable and there is more than one explanation. Explanation is then closer to a conditional probability. Again, in the context of our simple example: the probability of muscle pain and the feeling of tiredness after a marathon is high. The marathon explains why person X feels pain and is tired. Here, explanation is more like a conditional probability and the value of the explanation depends on known probabilities and the match with the schema.

The classic view is that explanation is a deductive argument including background knowledge and inference rules such as modus ponens. The inference rules allow the sequential application of "if-then-else" statements in order to justify an explanatory target. If no detailed background knowledge is available, explanatory schemata or probabilistic rules can be used. Nevertheless, additional forms of explanation are possible.

Let's shift again to a practical example. Every teacher knows that at times it is difficult to introduce a new theoretical concept. Even with the best of efforts, it may be difficult for the students to grasp the theoretical elements that are being introduced. A single example, however, may result in an "aha" experience and complete understanding of the new material. Examples have explanatory value and can be most useful, in particular if they are typical or even prototypical (Diederich 2008a). An analogy requires the existence of a memory system to store and search examples which have varying degrees of similarity. In summary, the following processes can generate explanations acceptable to humans:

- Logical deduction by use of inference rules
- Probabilistic rules including conditional probabilities
- Schemas based on the similarity between cases
- The provision of one or more example that is typical or even prototypical.

In artificial intelligence, several of these processes are often combined to arrive at systems that either use or generate explanations.

3.5.1 How and Why Explanations

As indicated earlier, it is useful to distinguish between two general types of explanations, e.g. causal explanations that are acceptable answers to "why" questions in contrast to the step-wise explanations that are acceptable responses to "how" questions.

A "how" explanation consists of a set of rules that map a given input to an output. In explanation-based generalization, a sequence of rules explains how the single training example is a positive instance of the target concept. "Why" explanations are typically used in conversational agents. After any question or statement, the user has the option to ask "why" and the system will justify asking that particular question or making a certain statement at this point in time. In this sense, a "why" explanation does not require deep causal reasoning or the identification of a limited set of inputs that causes some output. A justification for a question or a response to a user is given.

3.5.2 Generating or Identifying the Best Explanation

Very often, more than one explanation is possible and explanations may even compete. Thagard and Litt (2008) identify the evaluation of competing explanations as one of three major processes if an AI system aims to simulate human reasoning. The term "abduction" describes the inference to the best explanation as well as the generation of hypotheses (Thagard and Litt 2008). There are three criteria for selecting the best explanation:

Consilience: how much does a hypothesis explain?

Simplicity: how many additional assumptions are required to carry out an explanation?

Analogy: Are there hypotheses whose explanations are analogous to accepted ones?

3.5.3 Explanation and the "Theory of Mind"

As stated above, experienced teachers find it easy to adjust explanations for an audience. Teachers can do this because they have a very good Theory of Mind (TOM) of what a listener can understand. This Theory of Mind capability allows a teacher to pick the right explanations including demonstrations and examples. Any advanced form of artificial intelligence must have a very good theory of the human mind otherwise no good explanations can be offered. In the chapter on risk to artificial intelligence, examples will be given how a weak theory of mind can significantly reduce the ability to communicate in artificial intelligence systems and humans.

How are TOM models realised in an AI system? Shvo et al. (2020, pp. 3–4) identify a set of desiderata for logic-based TOM models in explanation:

multi-agent: … a multi-agent setting to support representation of the beliefs of one or more explainer and explainee.

agent-type agnostic: … different agent types whose beliefs may be internally represented, inspectable, and revisable in diverse ways.

belief based: … model the possibly false or simply incomplete beliefs of explainers and explainees.

reason about the beliefs of others: … must allow an explainer to reason about the explainee's beliefs when providing the latter with an explanation since, due to their possibly differing beliefs, an explanation for the explainer may not be an explanation for the explainee.

support belief revision: … must enable the explainer to consider how an explanation is assimilated by the explainee, and in particular how the latter revises their beliefs given potential explanations which may be inconsistent with their current beliefs.

explanations can refer to beliefs: … must allow for explanations that themselves refer to beliefs.

While such a functionality is impressive, it is difficult to believe that a purely logical system that reasons over states of beliefs can model the cognitive processes in humans when it comes to understanding and relating to one another. Even if the state and beliefs are inconsistent, contradictory and conflicting, and assuming that a system such as the one outlined by Shvo et al. (2020) can appropriately capture these various states and beliefs, this still falls short of coming close to a human being with a multitude of motives and needs. Human cognitive processes are connected and integrated with needs and desires, in addition to a memory system that captures a vast amount of knowledge and a lifetime of experience.

3.6 Explanation and Black Box Machine Learning

At the time this book is written deep learning is at its peak of popularity. Deep learning systems are artificial neural networks with multiple hidden layers. Artificial neural networks learn by associating input patterns with target output vectors. After learning has been completed, even an input pattern that is partially complete can generate the right output, or one that is very similar to the target vector used during training.

A neural network can map input vectors to outputs even if there is no similarity among the inputs at all. This is typically illustrated by use of the "exclusive-or" (XOR) problem: Two input vectors with two elements each and different values for each of the variables can be successfully associated with the right output if there is a hidden layer between the input and output units. Deep learning systems can utilise multiple hidden layers in order to achieve high accuracy in predicting outputs. Machine learning researchers and practitioners put high value on the generalisation capability of the neural networks. The system should perform well even if inputs are presented that have not been used during the learning period and even if these inputs are partially distorted or incomplete. Deep learning systems and other machine learning applications such as support vector machines can achieve this performance even if input and outputs are high-dimensional. Frequently in text and image processing, input and output vectors are large and sparse, i.e. only few input variables have non-zero values.

The current boom of machine learning systems and particular deep learning applications rests on the high accuracy these techniques can achieve as well as their robustness and generalisation capability. However, this performance comes at a cost. Black box machine learning systems cannot naturally explain how a certain result was achieved or why a certain output was produced (Diederich et al. 2009). Furthermore, there is no easy way to engage in a dialogue with a black box machine learning system to obtain explanations on how the system works internally. Symbolic artificial intelligence systems have an advantage here, and over the last decades, many researchers have tried to build explanation capable machine learning systems that can generate similar explanations to rule-based symbolic AI systems.

Machine learning researchers who worked on explanation systems often adopted performance criteria from symbolic artificial intelligence. The general idea is to extract rule sets from black box machine learning system such as artificial neural networks. These rule sets must satisfy a number of performance criteria. First of all, the behaviour of the rule set in recognising and processing new input patterns must be very similar to the black box learning system. In other words, the generalisation behaviour of the rule set and the underlying learning system should be near identical. This is the *fidelity* criterium: rule set and black box learning systems must behave in a similar way even for new or distorted inputs.

Another requirement is that the rule set is relatively small in terms of the number of antecedents that appear in the rules as well as the number of consequences or actions. It is generally believed that large rule sets do not offer explanation capability. Since humans have limited attention and working memory capability, it helps to have small and concise rule sets. In many applications, this is difficult to achieve if data sets are large and if there is a lot of noise as the result of missing or imprecise measurements.

It is also important to note that no rule set and no learning system would have any explanation capability if features cannot be interpreted by humans. The pixel that occurs in an image has no meaning in isolation. It has meaning in the context of surrounding pixels or the entire image. In a similar vein, a single letter or simple word that occurs in a sentence or text has little meaning in isolation; it is again context that provides most of the information. Deep learning systems offer an amazing performance, yet individual variables or features in patterns have little or no intrinsic meaning. This is literally not required for machine learning. It is, however, a requirement for human comprehension. In many applications, this means that the data sets used for deep learning must be comprehensible to some extent so that rules extracted from black box machine learning systems have meaning to humans. Or different data sets are used for explanation capability.

Frey (2019) reports on advancements of Google's Explainable AI systems and writes

> Explanations quantifies each data factor's contribution to the output of a machine learning model. These summaries help enterprises understand why the model made the decisions it did … any explanation method has limitations. For one, AI Explanations reflect the patterns the model found in the data, but they don't reveal any fundamental relationships in your data sample, population, or application.

Google describes the limitation of explainable AI systems in more detail (21 November 2019):

> Each attribution only shows how much the feature affected the prediction for that particular example. A single attribution might not reflect the overall behaviour of the model.

Attributions refers to features or sets of features and they depend on the model (the trained machine learning system) and the data set.

3.6.1 *Rule Extraction from Black-Box Machine Learning Systems*

Andrews et al. (1995) discuss the motivation behind rule extraction from black box machine learning systems. The five points below are still relevant for the current effort to extraction comprehensible descriptions from black-box machine learning systems. A brief review of Andrews et al. (1995) arguments is provided below.

3.6.1.1 Provision of a 'User Explanation' Capability

In symbolic artificial intelligence (AI), the term 'explanation' refers to an explicit structure which can be used internally for reasoning and learning, and externally for the explanation of results to a user (Diederich 2008a). Users of symbolic AI systems benefit from an explicit declarative representation of knowledge (for instance, a proposition such as "the car is red") and traditionally, symbolic AI systems are deductive in nature: Reasoning (including classification) is from the "generic" (expressed in the form of general rules) to the "specific" (an instance or individual that is to be classified). Even learning, if it is based on a large amount of background knowledge, is deductive in symbolic AI systems.

The explanation capability of a symbolic AI system is based on the intermediate steps of the reasoning process, e.g. a trace of rule firings or a proof structure, which can be used to answer 'How' questions. Gallant (1988) observes that the benefits of an explanation capability include a check on the internal logic of the system as well as enabling a novice user to gain insights into the problem at hand.

An explanation capability is considered to be one of the most important functions provided by symbolic AI systems (Diederich 2008a). The ability to generate even limited explanations is essential for the user-acceptance of such systems (Davis et al. 1977). In contrast to symbolic AI systems, neural networks have no explicit declarative knowledge representation, since traditionally, neural networks do not perform deduction (they are statistical learning techniques). Therefore, neural networks have considerable difficulty in generating explanation structures and the situation is no different in deep learning systems.

Practitioners in the field of symbolic AI have experimented with various forms of user explanation, in particular rule traces. However, it is obvious that explanations based on rule traces are too rigid and inflexible (Gilbert 1989) because rules may differ in their usefulness for the user. In addition, rule traces always reflect the current structure of a knowledge base. Furthermore, rule traces may have references to internal procedures (e.g. calculations); might include repetitions (e.g. if an inference was made more than once); and the granularity of the explanation is often inappropriate (Gilbert 1989; Andrews et al. 1995; Diederich 2008a, b). A clear lesson from the use of rule traces is that the transparency of an explanation is by no means guaranteed. For example, an explanation based on rule traces from a poorly organised rule base with perhaps hundreds of premises per rule is not comprehensible at all. Early rule extraction from neural network techniques (in particular those that are learning-based) generated a large number of rules with many antecedents (Diederich 2008a, b). Similar to the extended explanations based on rule traces in symbolic AI, the very large rule sets extracted from neural networks offer limited or no explanation capability.

Additional examples of the limitations of explanation capabilities in symbolic AI systems are described in Moore and Swartout (1989). In the field of expert systems, practitioners have been associating "canned text" with rules and instead of providing the user directly with the trace of rules, the sequence of pre-prepared texts is delivered instead. Again, this type of user explanation is very rigid; systems always interpret questions in the same way, there are no alternative response strategies. Although efforts have been made to take advantage of natural-language dialogues including mixed initiatives, user-models and explicitly planned explanation strategies, there is little doubt that these early systems were inflexible and rigid (Andrews et al. 1995;

Diederich 2008a). While the integration of an explanation capability (via rule extraction) within a trained neural network is crucial for user acceptance, such systems must avoid the problems already encountered in symbolic AI.

3.6.1.2 Transparency

While 'user explanation' is the primary objective for extracting rules from black box machine learning systems, the provision of 'transparency' of the internal states of a system is a close second. Transparency means that internal states of the machine learning system are both accessible and can be interpreted unambiguously. Such transparency would allow the exploration of regions in generalisation space which may lead to erroneous or sub-optimal decisions. Such a capability is mandatory if neural network-based solutions are to be accepted into 'safety-critical' problem domains such as air traffic control, the operation of power plants, medical surgery etc. Rule extraction offers the potential for providing such a capability (Andrews et al. 1995; Diederich 2008a).

3.6.1.3 Software Verification

If a black box machine learning application should be part of a larger software system which needs to be verified, then clearly this requirement must be extended to all components, including machine learning. Currently, rule extraction algorithms do not allow verification, i.e. they do not prove that a machine learning system behaves according to some specification. However, rule extraction algorithms provide a mechanism for either partially or completely 'decompiling' a black box system. This is about half-way to software verification because it allows for a comparison between the extracted rules and the specification (Diederich 2008a).

3.6.1.4 Improving Generalisation

If a limited or unrepresentative data set has been used in the black box training process, it is difficult to determine if and when generalisation fails for specific cases, even with evaluation methods such as cross-validation or a "What-If-Tool". By expressing learned knowledge as a set of rules, an experienced user can anticipate or predict a generalisation failure (Andrews et al. 1995). It may also be possible to identify regions in input space which are not represented sufficiently in the data set and need to be supplemented (Andrews et al. 1995; Diederich 2008a).

3.6.1.5 Data Exploration and the Induction of Scientific Theories

This has been one of the original objectives for rule extraction from black box machine learning systems and is essential for data mining and knowledge discovery. As Craven and Shavlik (1994) write 'a (learning) system may discover salient features in the input data whose importance was not previously recognised.' If a deep learning system has learned important and possibly non-linear relationships, then these are encoded incomprehensibly as weight vectors and additional parameters. Within the context of discovering new relationships, rule extraction algorithms significantly enhance the data mining capabilities of black box systems.

3.6.1.6 Translucency and Rule Quality

The core concept here is the perception (or "view") of the underlying black box system which is either explicitly or implicitly assumed within the rule extraction technique. "Translucency" describes the degree to which the internal representation of the black box system is accessible to the rule extraction technique (Andrews et al. 1995; Tickle et al. 1998; Diederich 2008a). More broadly, a taxonomy for rule extraction from neural networks has been introduced (Andrews et al. 1995; Tickle et al. 1998) which includes five evaluation criteria: translucency, rule quality, expressive power, portability and algorithmic complexity.

The following section describes the first two of the five evaluation criteria for rule extraction from neural networks (Andrews et al. 1995; Tickle et al. 1998) as these are most important for practical purposes.

The Simple Neural Network Case

Translucency describes the degree to which the internal representation of the black box system is accessible to the rule extraction technique. At one end of the translucency spectrum are those rule extraction techniques which view the underlying black box system at the maximum level of granularity i.e. as a set of discrete hidden and output units. Craven and Shavlik (1994) categorized such techniques as "decompositional". The basic strategy of such decompositional techniques is to extract rules at the level of each individual hidden and output unit within the trained ANN. In general, decompositional rule extraction techniques incorporate some form of analysis of the weight vector and associated bias (threshold) of each unit in the trained black box system. Then, by treating each unit in the ANN as an isolated entity, decompositional techniques initially generate rules in which the antecedents and consequents are expressed in terms *which are local to the unit from which they are derived.* A process of aggregation is then required to transform these local rules into a composite rule base for the ANN as a whole (Diederich 2008a; Tickle et al. 1998).

In contrast to the decompositional approaches, the strategy of the pedagogical (or learning-based) approaches is to view the trained ANN at the minimum possible level of granularity i.e. as a single entity or alternatively as a “black box”. The focus is on finding rules that map the ANN inputs (i.e. the attribute/value pairs from the problem domain) directly to outputs (Tickle et al. 1998). In addition to these two main categories, Andrews et al. (1995) proposed a third category which they labeled as “eclectic” to accommodate those rule extraction techniques which incorporate elements of both the decompositional and pedagogical approaches.

A number of authors have studied the algorithmic complexity of extracting rules from feedforward neural network. With regard to decompositional approaches, the basic process of searching for subsets of rules at the level of each (hidden and output) unit is exponential in the number of inputs to the node. Hence, heuristics are used to limit the depth to which the space is explored. Golea (1996) showed that extracting the minimum DNF (disjunctive normal form) expression from a trained feedforward net is hard in the worst case. Furthermore, Golea (1996) showed that the Craven and Shavlik (1994) algorithm is not polynomial in the worst case. This result does not apply to single-layer networks; however, extracting the best N-of-M rule from a single-layer network is again hard (N-of-M rule: if (N of M antecedents are true) then …).

Rule extraction from neural networks adopted criteria for the quality of the extracted rules. The set of criteria for evaluating rule quality includes (Andrews et al. 1995): accuracy, fidelity, consistency and comprehensibility of the extracted rules.

A rule set is considered to be accurate if it can correctly classify a set of previously unseen examples from the problem domain (Tickle et al. 1998). Similarly, a rule set is considered to display a high level of fidelity if it can completely mimic the behavior of the neural network from which it was extracted. An extracted rule set is deemed to be consistent if, under differing training sessions, the neural network generates rule sets which produce the same classifications of unseen examples. Finally, the comprehensibility of a rule set is determined by measuring the size of the rule set (in terms of the number of rules) and the number of antecedents per rule (Diederich 2008a; Tickle et al. 1998).

3.6.2 *Rule-Extraction for Whom?*

Assume for a moment a rule extraction technique has been developed or selected and has been applied to black box machine learning system. Let’s assume further that the quality criteria have been satisfied. A relatively small rule set has been extracted and the number of antecedents and consequences per rule is limited. To simplify the case further, let’s assume that the features that occur in the rules are human comprehensible.

Even if all these criteria are met, who would be satisfied and have full confidence in the performance of the black box machine learning system? Rule sets based on formal logic, even propositional logic, can be read by experts only. There would be no information whatsoever for a person without training in artificial intelligence or information technology. This is information for a select few, including experts of AI in information technology with a significant amount of knowledge of the application domain. There is no rule extraction technique or other technology that serves a layperson. But in a world with an artificial superintelligence, it is exactly the layperson who must have confidence. Hence, new criteria for explanation from black box machine learning systems are required, including those that serve a layperson and people with an intellectual disability.

3.7 Visualisation and What-If Explanation

Explanation can be based on language and logic, as outline above, but visualisation is an important tool as well. Visualisation is mostly used by the developers of machine learning models to see how changes in the input affect outputs or how information is represented inside the model.

Currently, "What-If" models allow the inspection of machine learning systems. Google's What-If is an interactive visual interface designed to probe and test machine learning models. It is an attempt to examine, evaluate and compare machine learning systems. Up to two models for a data set can be investigated as well as sensitivity to a range of features, sensitivity to optimisation strategies and the manipulation of individual inputs. This tool allows the comparison of multiple machine learning models within the same workflow as well as the visualisation of inference results and feature attributions. It is also possible to explore the similarity of data points.

However, the objectives of the visualisation tool are much broader and include what users would consider as fairness. Weinberger (2019) states

> There is currently — and rightfully — a sharp, global focus on making sure that AI doesn't perpetuate or, worse, exacerbate the unfairness of existing social systems. There are well-known cases where this has happened, and there's no doubt that cases are currently slipping past us all.

Obviously, there are different forms of fairness and the What-If tool includes a number of ways to explore these different types based on mathematical measures. They are optimisation strategies which include demographic parity, equal opportunity, equal accuracy and others. The objective is to optimise machine learning models with regard to these criteria but obviously, there are external factors as well. Fairness cannot be captured by a tool like What-If in its totality. As a matter of fact, the last judgement about fairness has to be made by humans.

While Google's What-If requires minimal coding, and optimisation strategies can be selected by the click of a button, it needs to be stressed that this is a tool for software developers. It is not a tool for laypeople. It is an instrument to explore machine learning models during development; it is not a tool to explain the behaviour of an artificial intelligence system that may exist anywhere in the world.

3.8 New Forms of Explanation

The discussion above has been quite conservative insofar as forms of explanation have been considered that are currently used in artificial intelligence. An advanced form of artificial intelligence requires advanced forms of explanation. Previously, we have considered asking "why" and "how" questions at certain points during an interaction with an AI system. The use of schemata, analogy and meaningful examples have been discussed as well.

In a scenario where one or more advanced forms of artificial intelligence are available for interaction and are changing the living environment of individuals, other forms of explanations have to be considered. First of all, we have to differentiate between an AI system explaining the way it operates from unexplained changes in the environment. Robotic systems, for example, may cause changes that are surprising and difficult to understand.

By way of example, let's assume you have a robotic gardener who works on weekdays while you are away. You return home and some very unexpected changes have happened and your robotic gardener is nowhere to be seen. If you want to have an explanation for what happened while you were away, there are a number of options: Firstly, contact the robot and ask what happened, assuming the robot is responsible for the changes. Or alternatively, assume that some other agent is responsible for the new developments. In this case, it should be possible to make a video of your garden including all the changes and obtain a written or spoken explanation of what happened while you were away.

We can immediately generalise and ask for a technology that takes almost any video as input and provides meaningful explanations for actions or changes in the video. Or we go even one step further and ask for some voice over that explains and interprets the world for us.

Wearable devices may be a part of solution to the problem of explaining unexpected changes in the environment. Assume you have a device that records everything you hear and see throughout the day. An AI system would quickly learn those aspects of the environment that do not change quickly: for instance, the furniture in your living room and kitchen. The system would easily notice changes, so the problem is one of attributing changes to a probable cause. Again, this can be framed as a learning problem.

As plausible explanations become available, a user with a "chatbot in the ear" would receive these ongoing explanations about changes. If an AI system has modified the environment, then it should be possible for the user to contact that system and to obtain information about the modus operandi and the reason for the changes. In this scenario, explanations are a constant voice over to make a world with an artificial superintelligence comprehensible to humans.

3.9 Explanation for Children

If an advanced artificial intelligence is human compatible, then it must be suitable for children as well as adults. Children learn about the world through observation and through social interactions with other children and with adults. This learning process requires a level of consistency; a world that is predictable and does not change at a rate that cannot be explained easily. Consistency is absolutely key. Not knowing what happens in the next hour or tomorrow has a damaging effect on the developing mind.

Attachment theory tells us that consistency in communication is of particular importance. A child who returns to a caregiver after some time away, and this may only be minutes, goes through a "reunion phase" during which positive attention and care is particularly important. When the child returns, it expects attention, positive emotions and expressions of love. Positive development is possible if the parent or caregiver pays a good amount of attention to the child during these reunion phases. However, if the caregiver is consistently distracted and does not behave predictably, then psychological disorders can develop. A negative or random response during the reunion phase is particularly problematic. In this case, psychological problems such as a reactive attachment disorder can develop. Children require a positive, consistent and loving environment and they do require all the explanation that is necessary to understand the world.

This point is crucial for the discussion here: A world with a safe artificial superintelligence must be safe for children. It is simply not sufficient if a few well-educated experts understand the world because they understand artificial intelligence. Whatever the future holds, it must be predictable and explainable to everybody, including children and people with intellectual disability.

Children need explanations and they ask for explanations as they develop mental models of the world. An advanced and safe artificial intelligence must be capable of providing explanations to children at various stages of development. The cornerstone of a human compatible and controllable artificial intelligence is the safety of children.

3.9.1 Satisfying Explanations for Children

Frazier's et al. (2016) research with preschool children shows that explanations are important to them in that they actively seek explanations from adults. Four- and five-year-old children were dissatisfied when no explanation was received to explanatory requests (Frazier et al. 2016, p. 718),

> but satisfied when receiving explanations, and their satisfaction varied appropriately across several levels of explanatory information.

Children typically failed to recall non-explanations and their recall of explanatory information was consistently high. Frazier et al. (2016, p. 718) emphasise that these results confirm that children not only actively seek informative explanations in their everyday conversational interactions with adults, they selectively retain the answers they receive.

3.10 Irrational Explanations

Miller (2019) argues that the field of explanation and AI should take inspiration from the way humans explain the world to each other. He focusses on "everyday explanations" that refer to particular facts (events, properties, decisions) rather than explanations of more general relationships, e.g. those in the scientific literature. He arrives at four conclusions about explanations (Browne and Swift 2020; Miller 2019):

1. Explanations are contrastive; that is, they "explain the cause of an event relative to some other event that did not occur".
2. Explanations are selective; that is, there is no explanation of a "complete" cause of an event.
3. Explanations are social; that is, they are presented relative to who the explainee is, and what they can be expected to understand.
4. Probabilities do not matter; statistical explanations of events are unsatisfying unless accompanied by causal explanations.

Explanations are also irrational, and frequently, these explanations are resistant to change. Irrational explanations are common in the daily practice of clinical psychologist. For example, a person may have come to the conclusion that their partner is a narcissist. As a result, all past and current behaviours are explained in terms of narcissism. Perception may be biased to the extent that only behaviour that is consistent with the hypothesis is memorised and other events are completely discarded. Irrespective of the true personality of the partner, the belief in narcissism is a sufficient explanation of the behaviour of the partner. Hence, alternative explanations of behaviour are frequently rejected.

The chapter on digital clones referred to unhelpful thinking styles in clinical depression. This includes cognitive dysfunctions such as personalization and catastrophising. A statement consistent with personalisation would be "everything is my fault" and in catastrophising, even small events are blown out of proportion. Nevertheless, these thinking styles completely determine the way a person sees the world. As a result, a person with an unhelpful thinking style such as personalisation will easily accept explanations that put blame on oneself. No matter how irrational these everyday explanations may be, they are accepted above more adjusted and logical conclusions.

It is important to note again that in western countries, a quarter or more of the population suffers from anxiety and depression at any given point in time. As outlined in the chapter on digital clones, these conditions have an impact on communication, and as a result, acceptable explanations of observations and facts. Irrational explanations are not necessarily the exception; they may be the norm.

A very advanced artificial intelligence system must have theory of mind models that account for clinically relevant explanations that a person may have. In other words, an artificial superintelligence must have theory of mind models that include all components of human development and psychopathology.

References

Andrews, R., Diederich, J., Tickle, A.B.: A survey and critique of techniques for extracting rules from trained artificial neural networks. Knowl. Based Syst. **8**, 373–389 (1995)

Browne, K., Swift, B.: Semantics and explanation: why counterfactual explanations produce adversarial examples in deep neural networks. https://arxiv.org/abs/2012.10076 (2020)

Craven, M., Shavlik, J.: Using sampling and queries to extract rules from trained neural networks. In: Proceedings of the 11th International Conference on Machine Learning, pp. 37–45 (1994)

Davis, R., Buchanan, B.G., Shortliffe, E.: Production rules as a representation for a knowledge-based consultation program. Artif. Intell. **8**(1), 15–45 (1977)

Diederich, J.: Instruction and high-level learning in connectionist networks. Connection Science. J. Neural Comput. Artif. Intell. Cogn. Res **1**(2), 163–182 (1989)

Diederich, J.: Explanation and artificial neural networks. Int. J. Man Mach. Stud. **37**, 335–357 (1992)

Diederich, J. (ed.): Rule Extraction from Support Vector Machines. Series: Studies in Computational Intelligence, vol. 80. Springer Verlag, Berlin, Heidelberg, New York (2008a)

Diederich, J.: Rule extraction from support vector machines—an introduction. In: Diederich, J. (ed.) Rule-Extraction from Support Vector Machines. Springer Verlag, Berlin, Heidelberg, New York (2008b)

Diederich, J., Tickle, A.B., Geva, S.: Rule extraction from neural networks: recent developments. In: Koronacki, J., Wirzchon, S.T. Ras, Z., Kacprzyk, J. (eds.) Recent Advances in Machine Learning. Dedicated to the memory of Ryszard S. Michalski. Springer Verlag, Berlin, Heidelberg, New York (2009)

Frazier, B.N., Gelman, S.A., Wellman, H.M.: Young children prefer and remember satisfying explanations. Cogn. Dev. **17**(5), 718–736 (2016)

Frey, T.: Increasing transparency with Google Cloud Explainable AI. https://cloud.google.com/blog/products/ai-machine-learning/google-cloud-ai-explanations-to-increase-fairness-responsibility-and-trust. 25 Nov 2019

Gallant, S.: Connectionist expert systems. Commun. ACM **13**(2), 152–169 (1988)

Gilbert, N.: Explanation and dialogue. Knowl. Eng. Rev. **4**(3), 235–247 (1989)

Golea, M.: On the complexity of rule extraction from neural networks and network querying. In: Proceedings of the Rule Extraction From Trained Artificial Neural Networks Workshop, Society For the Study of Artificial Intelligence and Simulation of Behavior Workshop Series (AISB'96) University of Sussex, Brighton, UK, pp. 51–59 (April 1996)

Hempel, C.G., Oppenheim, P.: Studies in the logic of explanation. Philos. Sci. **XV**, 135–175 (1948)

https://cloud.google.com/ml-engine/docs/ai-explanations/limitations. 25 Nov 2019

Köhler, W.: Intelligenzprüfungen an Menschenaffen. Spring Verlag, Berlin, Heidelberg, New York (1917)

Miller, T.: Explanation in artificial intelligence: insights from the social sciences. Artif. Intell. **267**, 1–38 (2019)

Moore, J.D., Swartout, W.R.: A reactive approach to explanation. In: IJCAI-89 International Joint Conference on Artificial Intelligence, pp. 1504–1510 (1989)

Shvo, M., Klassen, T.Q., McIlraith, S.A.: Towards the role of theory of mind in explanation. Chapter. July 2020. https://doi.org/10.1007/978-3-030-51924-7_5

Thagard, P.R.: The best explanation: criteria for theory choice. J. Philos. **75**(2), 76–92 (1978)

Thagard, P., Litt, A.: Models of scientific explanation. In: Sun, R. (ed.) The Cambridge Handbook of Computational Cognitive Modeling. Cambridge University Press, Cambridge (2008)

Tickle, A.B., Andrews, R., Golea, M., Diederich, J.: The truth will come to light: directions and challenges in extracting the knowledge embedded within trained artificial neural networks. IEEE Trans. Neural Networks **9**(6), 1057–1068 (1998)

Weinberger, D.: Playing with AI Fairness. https://pair-code.github.io/what-if-tool/ai-fairness.html. 25 Nov 2019

Chapter 4
Transhumanism

This chapter introduces the concept of transhumanism and its roots in history. A number of challenges to the transhumanist program are outlined next. Assuming individuals have freedom of choice in selecting physical or cognitive enhancements, the transhumanist program implies a *convergence of abilities* for any functioning society. The conditions for such a convergence are explored in detail. Next, psychological theories of self and personality are introduced and the question is asked to what extend transhumanist programs allow for the continued existence of the human self and aspects of personality. The role of the motivational system and the importance of its preservation through cycles of enhancement is outlined. Finally, the psychological implications of an infinite lifespan ("immortality") are explored, in particular with reference to the challenge to maintain a notion of self and personality.

4.1 Introduction

The concept of *transhumanism* refers to the enhancement of human functioning by chemical, biological and/or technical means. The first appearance of the term is attributed to Julian Huxley (1927; in Bostrum 2005, p. 7);

> The human species can, if it wishes, transcend itself – not just sporadically, an individual here in one way, an individual there in another way – but in its entirety, as humanity. We need a name for this new belief. Perhaps *transhumanism* will serve: man remaining man, but transcending himself, by realizing new possibilities of and for his human nature.

The Transhumanist Declaration, Version of March 2009 (in Bostrum 2005, Appendix) states that

> Humanity stands to be profoundly affected by science and technology in the future. We envision the possibility of broadening human potential by overcoming aging, cognitive shortcomings, involuntary suffering, and our confinement to planet Earth.

J. Diederich, *The Psychology of Artificial Superintelligence*,
Cognitive Systems Monographs 42,
https://doi.org/10.1007/978-3-030-71842-8_4

The declaration further states that there are

> possible scenarios that lead to wonderful and exceedingly worthwhile enhanced human conditions.

The age of transhumanism started a very long time ago with the many inventions in the medical and pharmaceutical sector. Today, healthcare offers life-saving services throughout the human lifespan. The medical sector is one of the largest of the US and other Western economies, e.g. contributing one fifth of the gross domestic product of the United States. The vast majority of people depend on these services, which do not only enable life, but reduce human suffering.

Reproductive medicine is commonplace and the genetic engineering of the human code has started as well. Many, if not all of these efforts, enable life, prevent disease and expand lifespan. Research in the health sector allows a quality of life unknown to previous generations, and on this basis, the transhumanist program goes one step further. The idea is to expand the human body/mind system *even without medical need*. Since medical research has given us so much, why not embrace the next step so that humans can be in charge of their own evolution? From this viewpoint, any individual should enjoy the freedom to choose enhancements that improve the quality of life. More to the point, groups of individuals, if not societies, should be able to undergo cycles of improvement.

Obviously, humans have enhanced their condition since the beginning of time. This includes the selection of herbs and food for nutrition and to cure ailments. During human prehistory, our ancestors drilled holes into skulls as a first primitive form of surgery ("Trepanation"). The intention was not so much enhancement but rather a cure for pain, an injury or disease.

Every use of a tool or instrument is an enhancement of the human condition. Human evolution is unthinkable without the co-evolution of tools. These tools, in turn, result in human change, e.g. increased upper body strength to use a hammer, a sword or to wear amour. A more modern example is the enlargement of the hippocampus (a part of the brain responsible for spatial navigation) in taxi drivers prior to the arrival of GPS (Maguire et al. 2006), even though this observation is subject to ongoing debate (Weisberg et al. 2019).

There are numerous and very unfortunate examples throughout history that humans have been selected because of traits or conditions that offer an advantage to a group or population. Alternatively, the killing of humans has been used when there is an absence of desirable traits. From the use of eugenics in ancient Greece to the criminal experiments of the Nazis, which brought immense suffering to millions, human beings have been selected for their physical or intellectual traits to offer advantage to a group or population. The extinction programs of the Nazis are just one extreme case of the human attempt to enhance certain traits and conditions while at the same time trying to eliminate others.

The academic debate on Transhumanism typically includes two viewpoints (Bostrum 2005): Bioconservatism vs Transhumanism. Since Transhumanism has been characterised already, here is Bostrum's (2005, p. 23) summary of the bioconservative position:

> The different strands of contemporary bioconservatism can be traced to a multifarious set of origins: ancient notions of taboo; the Greek concept of hubris; the Romanticist view of nature; certain religious (anti-humanistic) interpretations of the concept of human dignity and of a God-given natural order; the Luddite workers' revolt against industrialization; Karl Marx's analysis of technology under capitalism; various Continental philosopher's critiques of technology, technocracy, and the rationalistic mindset that accompanies modern technoscience; foes of the military-industrial complex and multinational corporations; and objectors to the consumerist rat-race. The proposed remedies have ranged from machine-smashing (the original Luddites), to communist revolution (Marx), to buying "organic", to yoga (José Ortega y Gasset).

Bostrum's (2005) summary of bioconservatism is quick and brutal. The statement does not really appreciate that there is a core belief shared by all three monotheistic world religions: (a) *Imago Dei*, human beings are created in the image or spirit of God (Sarayi et al. 2019); (b) Humans resemble the creator possibly in appearance but more in aspects such as cognition and maybe rationality; (c) The understanding that humans are created in the image of God does not only link humans to a creator, but nature as well, since it is assumed that nature was also created by God.

A sharp departure from the traditional view of humanity shared by large world religions is the transhumanist view that humans should take charge of their own evolution, advance/change evolutionary patterns or initiate enhancements that have nothing to do with natural development. Transhumanists not only radically challenge what it means to be human, they also challenge the traditional understanding of nature and frame it as something that can be changed at will. Cole-Turner (2011, Chapter 6: The Hopeful Cyborg) writes

> The cyborg is a generator of "narratives of apprehension" about technology and human technological proclivity. It stands in contrast to many of the traditional ways in which the world is ordered, a disconcerting form that raises questions about human nature, human identity, the relationship between the human and nonhuman in the world, and in particular, how to live wisely and wholesomely in a world constantly being reshaped by technology. As such, the cyborg encourages stories that capture both the anxiety engendered by the perceived negative effects of technology and a sense of wonder and awe at the power and scope of human technological agency. This is particularly true of questions that are concerned with the very essence of human personhood, about human nature, and the character of the relationship between human beings and the natural world.

As stated previously, any attempt to enhance human conditions at the individual or population-level is not new. The attempt to enhance human perception or cognition is also not new. However, an attempt to increase cognitive or perceptual abilities in order to match an artificial superintelligence, or to improve communication with such an entity, is a new development.

The conclusion of a study by the IEEE-FDC Symbiotic Autonomous Systems Initiative projects that by 2050 it will be difficult to separate living from non-living things (Sarocco 2019, p. 18). He highlights that enhancements will be the new normal:

> In the coming decades the relationship with the cyberspace will become seamless, a sort of sixth sense. This will create a gap between those who have it and those who don't. Having an implant seamlessly connecting to cyberspace may become a competitive advantage compared

to those who don't have it ... In turn, this will push people to require such implants, quickly leading to mass adoption.

4.2 Cognitive and Perceptual Enhancement

An example for a twenty-first century attempt to increase cognitive functioning is the use of stimulant medication, frequently without any medical need or prescription. The use and abuse of stimulant medication in college students is significant (White et al. 2006). Stimulants such as Ritalin and dexamphetamine increase attention and working memory for relatively brief periods of time (Arnsten 2006). Hence, the use of stimulants offers advantages to college students and anybody who wants to increase cognitive performance short-term. Amphetamines do not only improve attention and reduce fatigue in individuals who have deficits in this area, the German military used amphetamines to increase stamina and performance in soldiers during World War II. The use and abuse of the drug was so widespread that it received its own nickname: "Panzerschokolade" (tank chocolate).

Stimulants improve cognitive performance, and the medication is widespread and available on prescription or through illegal means. Individuals who are under the influence have comparative advantages in assessment and even in combat situations. This raises questions of fairness and equality, for instance in exams and battles. The equity and equality issues that arise from such enhancements represent significant challenges.

Obviously, it is possible to improve attention, working- and long-term memory by other means. For instance, a twenty-first century method to improve attention and working memory are brain training techniques (Bryck and Fisher 2012). These methods, often implemented in mobile apps, provide learning opportunities to selectively strengthen executive and memory functions. This represents an advancement by natural means rather than through the use of implants or chemical interventions. Hence, learning opportunities to improve attention and memory, for instance, are outside of the paradigm of transhumanism.

In the broad area of perception, corrections and enhancements of vision are commonplace, including the extension of visual abilities into the infra-red range to provide night vision. It is not inconceivable that night vision and other more advanced enhancements will become as commonplace as glasses are today.

Enhancement of smell may also become commonplace. Currently, animals with their superior sense of scent, are used in the security sector but it may also be possible to intensify the human sense of smell for the purpose of security, and also for far-ranging applications, such as food and wine production or the development of perfumes, cosmetics and aphrodisiacs. It is not only possible, but likely, that groups who regard these enhancements as important will push for the development of various abilities.

Although enhancements of memory are currently commonplace through devices such as mobile phone reminder systems for appointments or daily activities, these systems are external to the biological human organism. At present, additional memory systems that directly interface with humans remain part of science fiction.

This may be because it is difficult to see how other properties of the neural cognitive system can be enhanced, such as the processing speed of the brain, without replacement by electronic devices or other forms of technology. Clearly, stimulant medication has an effect but biological neurons are incredibly slow processing elements and there is simply no comparison to electronic switches or even optical computing.

For the sake of the argument, assume that cognitive functions typically covered by intelligence tests such as verbal comprehension, perceptual reasoning and working memory can be significantly enhanced by use of some chemical, biological or technical intervention (or combinations thereof). Let's assume a temporary or permanent boost of functioning as measured by the intelligence quotient (IQ) is possible and let's assume that the new mean IQ is 200 instead of the current 100. By comparison, Albert Einstein had an IQ of 160. An IQ of 200 is so extreme by today's standards that statistically it does not make sense, but again, we are talking about superhuman intelligence.[1]

If superintelligent transhumans are committed to equality and diversity, would they accept living in a society with individuals who have an IQ of 100? This would be similar to individuals with an IQ of 122 living in a society of individuals with an average IQ of 65 (i.e. an area that represents a mild mental disability)? Would it be fair to say that superintelligent humans would most likely associate with individuals of comparable intelligence, in particular for anything that entails creative work? And what might they create?

The argument can be extended to extreme enhancements of perceptual abilities. Can we reasonably assume that individuals with similar enhanced abilities in areas such as smell, vision and hearing would form identity groups and prefer to associate with others within their groups? And would such groups associate less with persons who have advanced abilities in completely different areas?

The point is that for any society of transhumanist individuals, there needs to be a "convergence of abilities". If this is not the case, a breakdown in communication and social cohesion is highly likely. Without a convergence, there are no groups with similar cognitive and perceptual abilities. So far, proponents of transhumanist thought have not reflected on the conditions for such a conversion and the impact on societies. It is likely that there needs to be an agreement on how to enhance, and what abilities to enhance, to enable some form of coordinated living.

[1]Bostrom (2014) is talking about an artificial superintelligence with an IQ above 6,000 and at the same time confirms that such a metric makes no sense for very intelligent machines.

4.3 The Convergence of Abilities

A convergence of technical developments is not unusual and can easily be demonstrated. By way of example, in 2020, two types of smartphones are widely used. One type from Apple, with the corresponding operating system, and the other from Google. Annual updates for the most popular phones are available, and as expected, software for the devices is updated more frequently than hardware.

As intended by the competing companies, most consumers stay in one hard- and software environment and update as frequently as budget and interests allow. Those who need technology for work, or people who have sufficient funds, update more frequently; less fortunate (or less interested) consumers may need (or want) to wait a few years.

Technological developments as well as market pressure have resulted in a convergence—a stable state, with two main competitors in an international market and two separate hard- and software environments. Software solutions that were once considered an alternative, such as Windows phones, have moved to the background. Since consumers are mostly happy with the environment they adopted, the relatively stable state will continue until there is a disruption through technology breakthroughs or market pressures.

Now let's assume the hard- and software is part of somebody's brain. On the hardware side, relative stability may be supported by the fact that frequent surgery or injection of implants may not be desirable. The consumer may also wish to stay in the same software environment, similar to how they remained with particular smartphones, if enhancements have an impact on conscious experience and if transitions to new software are difficult. Nevertheless, if enhancements truly result in an improved quality of life, then the consumer may wish to update as frequently as possible within a given environment and enhancement trajectory. For instance, someone who wishes to improve their intelligence or perceptual abilities may choose frequent enhancements to achieve better and better outcomes.

It is immediately transparent that all of this creates significant equity and equality issues. What if consumers cannot afford an update? Clearly, wealth on the one side and poverty on the other always had a direct impact on physical well-being and conscious experience. But here we are talking about something much more dramatic: enhancing the very thing that makes human experience possible.

When the human brain is involved, any equity and equality argument has *unparalleled strength.* When a billionaire and his friends make the first Space X starship trip to the moon, we are all happy to sit back, relax and watch. However, when electronic implants and brain-computer interfaces are used to repair the neural damage caused by a stroke, it is extremely important to make the technology as widely available as possible without any delay. Hence, the situation is identical to the availability of life-saving medication and vaccines.

These equality issues may result in the consideration of government support for enhancement and maybe *synchronisation of the timing* so that groups of individuals go through the experience simultaneously. In other words, governments might decide on the type of enhancements and availability directly. The situation is not unlike the coronavirus vaccinations that are available towards the end of 2020: Governments decide what type of vaccine is available and who will get it first.

4.3.1 Enhancements, Personality and the Notion of "Self"

It is necessary to introduce two psychological concepts at this point in time: personality and self. The psychological definition of personality is "stable patterns of behaviour over time." Personality develops during the first two decades of human development and is then relatively stable. It includes the things we prefer as well as our daily routines and typical responses to situations. The definition of self is more complicated. The APA Dictionary of Psychology defines self as

> the totality of the individual, consisting of all characteristic attributes, conscious and unconscious, mental and physical.

The notion of self is crucial for any discussion of the concept of transhumanism. This is because the enhancement of physical and mental functions serve as an improvement of self. The search for immortality, for instance, aims at extending the self into infinity. Without an improvement or extension of self of an individual, no biological, technical or mental enhancement makes sense. Transhumanism is fundamentally narcissistic since it aims at a preservation and temporal extension of the self at all cost. Even the more extreme versions of transhumanism that aim at uploading the human mind to a machine put highest value on the preservation of self. Cryonics (i.e. freezing a human corps with the hope that resuscitation may be possible in the future) is the promise of the re-emergence of the self at some point in time.

Nobody argues for a form of transhumanism that eliminates the notion of self because we cannot know what this means or how any experience is possible without self and consciousness. Hence, the impact of enhancements on the self and on personality is a very important question. Several theories of the self have been promoted throughout the centuries. Some are based on recent developments in the neurosciences and cognitive psychology, and hence, are relevant for the discussion here.

4.4 The Concept of Self

Before philosophy and psychology focused on the concept of self, the human soul was the main subject of the discourse. But since the human soul is not open to an empirical investigation, the discourse moved to topics such as the "true inner self", the appearance of an individual to others, and the external expression of the self. Is there something like a "true inner self"? The notion of self was not always at the core of psychological and philosophical discussions, however, since the second world war and possibly under the influence of psychoanalysis and humanistic psychology, researchers and practitioners have increasingly focused on the self, self-concepts and even self realisation.

In philosophy, René Descartes (1506–1650) and his "cogito, ergo sum" (I think, therefore I am) established the self not as an acting entity but as a thinking and reflecting individual (Mummendey 2002). Gottfried Wilhelm Leibniz (1646–1716) observed that we think of a substance when we say "me" or "I". The philosopher and Anglican Bishop George Berkeley (1685–1753) offers an approach that is very attractive to modern psychology. Berkeley believed that human identity is fundamentally based on perception, both the possibility to perceive but also being perceived by others. Is it possible to perceive oneself, and as a result, become aware of the existence of oneself?

David Hume (1711–1776) emphasises that human consciousness, or self, is not a substance but it consists of perceptions, the content of the self. The process of perception creates the impression of a self, which is nothing but perceptual content; impressions and their relationships.

William James (1842–1910) distinguishes between the "material me", a perception of the body, the home the family and familiar objects, and the "social me", the knowledge of others and their views and judgements. The third entity is the "spiritual me", the experience of thinking and emotion. According to James, the experience of cognitive activity is the direct experience of the human soul.

In Sigmund Freud's theory (1856–1939), the self is not a substance but an entity that negotiates between motives and drives on the one side and social or cultural requirements on the other. While a personality is commonly seen as a structured entity, it does not have to be static. A dynamic interpretation of personality is possible, which entails a collection of processes that are expressed in the interaction with others. Hans Thomae (1915–2001) identifies a number of processes, including a certain orientation or direction in life and a tendency to analyse events or facts deeply or at the surface.

As can be seen from this chronology of thinkers, the "self" is not an easy subject for any scientifically oriented psychological discourse. Roy Baumeister writes in his book on "the self in social psychology" (in Mummendey 2002, p. 11) and describes how the notion of self became a social problem, illustrating the issue by pointing to statements such as

> The self is not known directly but is either observed in action or inferred from social events which is why … it is harder to know and understand than say a table or spoon.
>
> The self is a member of groups or relationships.
>
> The self is involved in interacting with others.
>
> The self has an executive function. This enabled the self to make choices.

But does the self really exist? Thomas Metzinger (in Northoff 2013) argues that the self as a mental substance is an illusion. However, we do have "self models" as a result of experience that directly correspond to information processes in the brain. In other words, since we do not have direct access to this neural processing (i.e. the embodiment of information processing in the brain), we create self models. There is no other way for the mind to work in order to create a sense of coherent experience, since there is nothing but the massively parallel flow of information in the brain. The phenomenal self is simply the content of what is perceived at any given point in time. Hence, the assumption of the self as a mental entity results from a false positive, an erroneous inference from our experience (Northoff 2013, p. 2).

This definition of self is appealing to the transhumanist program. Metzinger's theory of self is attractive because, with every update or enhancement of the body/mind system, self models that directly relate to information processing in the brain adjust automatically. In Metzinger's theory, the self would never be challenged as a result of an enhancement.

However, this view of the self as nothing but perceptual content has problems. Children who are younger than two years old are perfectly capable of learning, however, practically no adult has childhood memories from the time between birth and age three. An accepted explanation is that young children do not have the self that allows them to integrate information and to distinguish their own person from that of the environment. Howe and Courage (1997, in Conway and Pleydell-Pearce 2000) argue that it is not until the development of the "cognitive self" at around 24 months of age that the infant is able to encode experiences in terms of self. The cognitive self emerges when the infant is able to *distinguish the I from the me,* that is, when the child is able to perceive the self as an object (Conway and Pleydell-Pearce 2000, p. 277). The I is a sense of the self as a thinking entity. A child younger than two years can learn but cannot organise knowledge in terms of a structured self-system (Conway and Pleydell-Pearce 2000, p. 277). Such organizational ability requires a notion of self to memorise personal experiences in a way that can be retrieved and retold.

4.4.1 Self Memory Systems

Klein (2012, p. 287) points out that "neuropsychological studies of self suggest that the singular self of everyday experience … is informed by a number of functionally separate neurocognitive systems" including

1. Episodic memories of life events.
2. Semantic representations of one's personality traits.
3. Semantic knowledge of facts about one's life.
4. A timeline of experiences that involves episodic memory.
5. A sense of personal agency and ownership.
6. The ability to self-reflect: To form meta-representations where the agent is the self, and make inferences on the basis of those representations.
7. The physical self: The ability to represent and recognize one's body.

Rogers et al. (1977, in Klein 2012, p. 289)

> found that judging trait adjectives for self-descriptiveness ("Describes you?") led to better recall than did structural (e.g., "Printed in capital letters?"), phonemic (e.g., "Rhymes with XXX?") or semantic (e.g., "Means the same as XXX?") encoding of the same material.

Memory is better if there is a reference to oneself.

In the Greenwald et al. (1989) experiments, subjects produced friends' names and then constructed sentences including that particular name together with an assigned (target) noun. Recall tests showed superior memory for target nouns used in sentences with own friends' (self-generated) names vs. nouns used in sentences with others' friends' names. Greenwald et al. (1989) point to *strong mnemonic effects associated with the self* and these can be understood in terms of familiar, ordinary memory processes. This is a very strong, important and consistent result from memory research: Material that is linked to the self, is important to oneself, and has emotional content, is remembered well. Hence, the self is important in everyday functioning since it has a strong impact on the retrieval of memory contents.

Martinelli et al. (2013) explored self-memory systems and the neural substrate of these. The key finding was that episodic autobiographical memory predominantly activates posterior and limbic regions including the hippocampus (relatively old parts of the brain and those important for learning). Semantic autobiographical memory is associated with anterior as well as posterior and limbic activations, but to a lesser degree than episodic autobiographical memory. The conceptual self mainly recruits medial prefrontal structures (typically associated with attention and working memory). Martinelli et al. (2013, p. 11) claim that these results demonstrate

> the existence of a subsystem within semantic memory that processes specifically self-related information based on trait-knowledge …. the declarative self is not monolithic but a multidimensional construct comprising distinct representations at different levels of abstraction.

4.4.2 The Self and Autopoiesis

The concept of autopoiesis (i.e. a system capable of reproducing and maintaining itself) is useful for a theory of the self. According to Bourgine and Stewart (2004, Abstract)

> An autopoietic system is a network of processes that produces the components that reproduce the network, and that also regulates the boundary conditions necessary for its ongoing existence as a network.

The concept of autopoiesis has its origin in the work of Maturana and Varela, two Chilean biologists. The concept is important for the discussion of self since an autopoietic system (1) includes the ability to reproduce itself and (2) creates boundaries that separate the system from the outside.

The standard model is the living cell surrounded by membranes and the ability to exchange substances and information with the environment. An autopoietic system includes processes that produce components used for the maintenance and reproduction of the system. This includes the boundary that separates the inside of the system from the outside.

According to Bourgine and Stewart (2004, p. 329), Maturana recently proposed the following definition of Autopoiesis

> A molecular autopoietic system is a closed network of molecular productions that recursively produce the same network of molecular productions that produced them and specify its boundary remaining open to the flow of matter through it.

Again following Bourgine and Stewart (2004, p. 329), Varela proposed a simpler definition by providing three basic criteria: A system is autopoietic if: (a) it has a semi-permeable boundary, (b) the boundary is produced from within the system, and (c) it encompasses reactions that regenerate the components of the system.

Bourgine and Stewart (2004, Abstract) also offer a definition of cognition:

> A system is cognitive if and only if sensory inputs serve to trigger actions in a specific way, so as to satisfy a viability constraint.

In other words, the system is cognitive if it wants to preserve its identity and survive.

Is the self an autopoietic system? Thinking includes self-attribution, the identification of preferences and behaviours that belong to oneself and not to others. For instance, I always wear a wide-brim hat in Australia plus clothing that offers UV protection. This is something *that I do*, other people make other choices. In addition, there is the self-attribution of personal memories, in particular episodic memories, as belonging to oneself; the acknowledgement of a personal history that is unique and belongs to one person only. There is a boundary surrounding the self, a system *without substance* that reproduces itself by a process of self attribution. The inside is formed by all the memories, behaviours, beliefs and preferences that belong to a person, that form an identity and a self. There are many other behaviours and preferences that can be attributed to other people, and while these may overlap, the totality of behaviours and beliefs as well as personal memories belong to the one person only, and the unique self of this individual.

As mentioned, it is impossible to remember the earliest events in life because the self has not yet been formed without the vital boundary between one's own experiences and unrelated events in the environment. Later in human development, the events of everyday life largely confirm to the person that which is known about the self (the self attribution) with occasional expectation failures resulting in new realisations.

Why is all of this important for discussion of transhumanism? Despite an openness to the many enhancements and updates that are promoted by the program of transhumanism, very few would be willing to abandon the notion of self. This is because a continuing mental existence, without an integration of experience and personal memory, would equal personal death and hence is not a desirable option. Consequently, the transhumanist program and the idea of enhancements that overcome the current human condition very much rely on a preservation of the self.

4.4.3 Enhancement and Social Cohesion

There is the obvious problem that, at the point in time the mind/body system is enhanced, the personal environment does *not* change. This includes social relationships. With an update of the body/mind system, the person who was intellectually and psychologically dependent a moment ago, may now be superior. However, this person's social relations may not adjust as quickly as required by enhancements.

From a human developmental perspective, upgrading skills and knowledge takes time, and during this period, an individual and the personal social network co-evolve. For instance, learning while attending university may take 3–4 years, and during this time, the learner's network of peers and family members participate in this journey of education. In other words, there is a coevolution of the student and his or her social network. By comparison, in the case of an enhancement, the improvement is immediate and the personal social network has no time to adjust.

Let's take the case of improvements and adjustments associated with current surgeries for cataract or hip replacements. The former provides better vision, the latter better movement. Currently, there still is a period of time after surgery for the patients' physical recovery, and this time is useful not only for the patient but also for the social network. Cataract surgery and hip replacements are human enhancements, but the resulting changes are unlikely to be so dramatic that partnerships or a family network suffers undue stress as a result of these enhancements. However, brain-machine interfaces offer significant enhancements that would be more immediate and would require more prolonged social adjustments.

4.5 Brain-Machine Interfaces

It is clearly beyond the scope of this book to review all possible brain-computer interfaces as well as possible implications for the concept of transhumanism. The focus here is on the technology developed by the Braingate project as well as on the "neural lace" as it offers connectivity between electronic implants and biological neurons on a large scale. As we have seen already, linking the brain with cyberspace is an important part of the transhumanist program.

The Braingate project (https://www.braingate.org) focuses on the therapy of neurological disorders and not on human augmentation or enhancement. Braingate is a collaboration of a number of universities and hospitals and includes work on assistive communication, movement restoration, neurotechnology and neurotherapeutics. The objective is to restore communication, mobility and to some extent independence in people with neurological diseases or injuries. This includes patients affected by stroke. The researchers implanted an array of microelectrodes into the brain of patients and demonstrated that neural signals indicating the intent to move or to communicate can be decoded by a computer in real time, and in turn, can be used to operate external devices. For instance, people can control a computer cursor simply by thinking about the movement, that is, a movement of their own paralysed hand or arm.

In addition, the Braingate project works on wireless medical technologies to record and monitor neural activity in order to diagnose and manage neurological diseases. Neural signals are recorded from the motor cortex using intracortical microelectrode arrays. These signals are then translated into point-and-click commands (Pandarinath et al. 2017).

The neural lace idea has its origin in fiction, an implantable "neural lace" is described by the Scottish novelist Iain M. Banks in *Look to Windward*, a part of his "Culture series" books (Sanford 2018). A neural lace can be injected into biological structures like the brain. Lattice-brain implants integrate into the brain and enable neuronal recordings. This approach goes as far as injecting electronics through a needle into neural tissue. The technique overcomes the immune-response problem of brain implants and allows measurements and the modulation of neural circuits. Once implanted, the brain can grow throughout the neural lace. Once injected, the two-dimensional mesh takes the form of a cylinder that gets filled with the tissue. The result is a tight integration of the neural and electronic network.

On 17 July 2019, Neuralink, a Californian company, introduced a scalable, high-bandwidth brain-machine interface (Musk, E. and Neuralink 2019). Neuralink built arrays of small and flexible electrode "threads", with as many as 3,072 electrodes per array distributed across 96 threads. The electrode array is part of a small implantable device that contains custom chips for low-power on-board amplification and digitization (Musk, E. and Neuralink 2019, Abstract).

In his presentation on 17 July 2019, Elon Musk said that ultimately he wants "to achieve a symbiosis with artificial intelligence." Even in a "benign scenario," humans would be "left behind." Musk wants to create technology that allows a "merging with AI." The goal is to read neural spikes from the brain, process the information locally, utilising an external device placed behind the ear of an individual and then to connect to mobile devices via Bluetooth. Note that the Neuralink technology allows both the recording of action potentials as well as the stimulation of neurons with the declared goal to create sensations.

Neuralink devices break René Descartes "cogito, ergo sum" (I think, therefore I am). *I think but the thoughts may not be mine.* Not surprisingly, the question and answer session of the Neuralink presentation on 17 July 2019 raised the question about "concept telepathy". If neural activation patterns can be read off the brain, and if information can be extracted from these patterns, which in turn correlates with what a person sees or feels at a given moment in time, then this information can be broadcast over the Internet. In other words, the perception of a person can be relayed to millions of people instantaneously. In turn, a person with a Neuralink device could be stimulated in relevant brain regions directly based on the broadcast of other individuals.

The quality of the sensation created by neural implants is very much in question. Nevertheless, this "telepathic" sharing of information is very much different to seeing a movie. This is about the direct stimulation of the brain. It's about instantaneously sharing perceptions, and potentially thoughts, and emotions. The processing device that creates thoughts and emotions, the brain, is now directly receiving information from other people's brains.

4.5.1 BMI, Telepathy and Psychosis

Psychosis is a mental disorder affecting thoughts and emotions as well as the loss of contact with reality. According to DSM-5,

> Hallucinations are perception like experiences that occur without an external stimulus. They are vivid and clear, with the full force and impact of normal perceptions and not under voluntary control.

Hallucinations may occur in any modality (sense), however, auditory hallucinations are most common in schizophrenia. BMI devices like the ones proposed by Neuralink may create perceptions without an external stimulus (i.e. no direct stimulus from the environment to the individual). The artificially induced perceptions do not directly correspond to objects or events in an individual's environment. However, these perceptions are not psychotic in nature because it is assumed they are under voluntary control. In the Neuralink scenario, a user would control these perceptions by use of a mobile app.

Again, the notion of self is critical here. If the externally generated information induces a thought disorder (similar to schizophrenia), then identity and self would disintegrate and no voluntary control would be possible any more. This may be one of the side effects of using cortical brain-machine interfaces—nobody knows at this point in time. One thing is certain, theories of self will be thoroughly tested in the near future by the availability of this type of transhumanistic enhancement.

Currently, many people are flooded by the thoughts and emotions of others through social media. Concerns about the mental health impact of social media are commonplace. This includes self harm in young people, often stimulated by images and videos. For the first time in human history, the thoughts and emotional expressions of people far away are instantaneously available. Now imagine this content being broadcast directly into the brain, including sexual experiences and violent content.

What difference is there between my own thoughts and those of others? What impact has this sharing of information on the notion of "self"? The framework of autopoiesis applied to the concept of self implies that there is a crucial difference between experiences and memories attributed to an individual (the "self") and those of others. How is it possible for a person to maintain a self when instantaneously flooded with the perceptions, thoughts and emotions of others? There is a real risk that devices such as the ones envisioned by Neuralink allow a connection between neural tissue and an artificial intelligence, however, it is not a link between a person and an AI.

Why is receiving a broadcast by use of a brain machine interface different from watching TV? Television, and particular live TV, certainly allows us to engage with events very far away, directly (the German word for TV "Fernseher" captures this very nicely: the TV is a "far seer"). Many TV shows and live broadcasts are designed so that the user is fully immersed. Live pictures from the moon or human suffering in conflict zones most certainly stir emotions.

Television, however, happens on a screen, and the viewer still has an awareness of their immediate surrounds. Even if a user is fully immersed in a computer game, the remaining awareness of the current surrounds cushion them from the experience of the game. Hence, the user is still in a position to terminate the game. Is this still the situation if the experiences of others are directly inserted into the brain?

4.6 Human Motivation

All psychological theories and frameworks include motivation as a core component of human behaviour. Motivation is key to the explanation of behaviour in psychoanalysis, because large parts of the theory are devoted to unconscious motives and drives. In psychoanalysis, unconscious psychological forces, including hidden desires, shape human behaviour. The theory is very general in the sense that it explains the selection of partners, consumer behaviour as well the contents of dreams. Since motives are unconscious, and needs can be suppressed, the empirical testing of the theory is difficult.

In Maslow's (1954) theory, motivation is the attempt to satisfy a number of basic needs. Maslow formulated a hierarchy of needs with physiological requirements such as air, food and sleep at the bottom (the most fundamental level), followed by safety requirements as well as cognitive and social needs. The latter result in positive self-esteem and psychological well being. Self actualisation and transcendence are at the top of the hierarchy and represent the so-called growth needs.

Another significant psychological framework that includes motivation as a cornerstone of human behaviour is learning theory. The focus is on the stimulus control of behaviour on the one side and response allocation on the other. Pavlovian conditioning is a good example of stimulus control: conditioned behaviour follows the presentation of a stimulus. Response allocation is the core of instrumental conditioning. Given the free flow of behaviour in every day life, how do outcomes and consequences play a role in the selection of a particular behaviour? The law of effect (Thorndike 1932) and the significance of consequences are the cornerstone of behaviourism.

The response allocation approach is a behavioural theory that assumes a very large range of activities is always available to an individual. This approach is particularly focused on the role of rewards in the distribution of behaviour or the selection of a particular response. It is obvious that response allocation is very valuable for the explanation of consumer behaviour, including the demand for certain products and services. From the viewpoint of clinical psychology, the "behavioural bliss point" is a particularly interesting concept. Giving a free flow of behaviour and no restrictions, the behaviours at the bliss point are stable and absolutely the preferred options of an individual. A person who is close to a behavioural bliss point demonstrates patterns of sometimes maladjusted behaviours and frequently treatment resistance.

In clinical practice, parents of children and adolescents often express concerns about excessive gaming behaviour, which is rather akin to behavioural bliss. A young person may spend many hours a day playing video games and thereby neglecting school work and social activities. When too many hours a day are spent on a particular computer game, concerns about addiction are frequently expressed by caregivers. The individual is completely immersed in the game which is so rewarding that other activities are neglected, and as a result, there is no reward opportunity for anything that is not game associated.

Behaviorism teaches that outcomes matter and the cognitive revolution in psychology has not challenged this very basic fact. Even within the broad area of political theory, for instance Marxism, the outcome of productive behaviour is essential and the result of work should not be divorced from the activity as such.

In other words, *we do something to get something*. If the connection between human activity and reward is weakened or eliminated, psychological and behavioral problems arise. The use of illicit drugs is one example: drugs like cocaine have an immediate impact on the brain's reward system. Frequent users are seeking the drug in order to obtain pleasure and reward without having to engage in behaviour that offers exceptional satisfaction without chemical stimulus.

Brain machine interfaces such as the one proposed by Neuralink risk to weaken or destroy the fundamental connection between behaviour and reward. In principle, an actor in the porn industry could use a neural lace to broadcast perceptions directly into the brains of millions of people. Given that sex is among the strongest reinforcers known to behavioural psychology, it is conceivable that some people may not want to do anything else but to receive this kind of satisfying input directly into the brain. In this scenario, neural laces are not just an enhancement but represent a significant solicitation.

4.7 The Extension of Life

Transhumanism is based on two propositions that are contradictory. The first assumption is that the human existence is imperfect at this point in time. As a result, human life should be modified by biological, chemical or technological means. The second proposition is that human life should be extended, in the extreme case, to immortality.

The question is, if life is so imperfect why should it be extended? The underlying assumption is, of course, an extended life by yet unknown enhancements and modifications would improve human existence to the point that it makes it worthwhile to live forever. It is unclear whether further modifications and extensions are required once a human life enters a stage that is deemed liveable. Is there an ongoing, never ending process of modification? Will this process of change accelerate with every enhancement, since improved technology will be available? Will the process stop if a stable state of subjective or objective quality has been achieved?

Once the program of transhumanism has been fully established, some members of society will embark on the road of enhancements and others will not. This is, of course, unless certain advancements become compulsory for everyone. In addition, we must assume that the process of enhancement would be slow for some and fast for others, unless the access to technology is government funded and regulated. This poses the question of the *rights of the unenhanced*. Will those who do not participate in a process of ongoing enhancement be in full possession of their civil and human rights or will they be reduced to a lower set of rights? Will there be direct or indirect pressure on those who are on the road of enhancement to accelerate the process in order to achieve the complete set of freedoms and rights that are afforded by society? The rights of the unenhanced are explored in the chapter on Neo-Luddism.

References

Bostrom, N.: Superintelligence: Paths, dangers, strategies. Oxford University Press, Oxford (2014)

Bostrum, N.: A history of transhumanist thought. J. Evol. Technol. **14**, 1 (2005)

Bourgine, P., Stewart, J.: Autopoiesis and cognition. Artif. Life **10**, 327–345 (2004)

Bryck, R.L., Fisher, P.A.: Training the Brain: Practical applications of neural plasticity from the intersection of cognitive Neuroscience, developmental psychology, and prevention science. Am. Psychol. **67**(2), 87–100 (2012)

Cole-Turner, R.: Transhumanism and Transcendence: Christian Hope in an Age of Technological Enhancement. Georgetown University Press (2011). Project MUSE muse.jhu.edu/book/10529

Conway, M.A., Pleydell-Pearce, C.W.: The construction of autobiographical memories in the self-memory system. Psychol. Rev. **107**(2), 261–288 (2000)

Greenwald, A.G., Mahzarin, R., Banaji, M.R.: Self as a memory system: powerful, but ordinary. J. Pers. Soc. Psychol. **57**(1), 41–54 (1989)

Klein, S.B.: Self, memory, and the self-reference effect: an examination of conceptual and methodological issues. Pers. Soc. Psychol. Rev. **16**(3), 283–300 (2012)

Maguire, E.A., Woollett, K., Spiers, H.J.: London taxi drivers and bus drivers: a structural MRI and neuropsychological analysis. Hippocampus **16**, 1091–1101 (2006)

Martinelli, P., Sperduti, M., Piolino, P.: Neural substrates of the self-memory system: New insights from a meta-analysis. Hum. Brain Mapp. **34**(7), 1515–1529 (2013)

Maslow, A.: Motivation and Personality. Harper & Row, New York (1954)

Mummendey, H.D.: Zur Substantialisierung des "Selbst"—Wurzeln und Erscheinungsformen. Bielefelder Arbeiten zur Sozialpsychologie, Nr. 202 (2002)

Musk, E., Neuralink: An integrated brain-machine interface platform with thousands of channels. White Paper. https://www.biorxiv.org/content/10.1101/703801v1 (2019). Retrieved 18 July 2019

Northoff, G.: Brain and self—a neurophilosophical account. Child Adolesc. Psychiatry Ment. Health **7**, 28 (2013)

Pandarinath, C., Nuyujukian, P., Blabe, C.H., Sorice, B.L., Saab, J., Willett, F.R., Hochberg, L.R, Shenoy, K.V., Henderson, J.M.: High performance communication by people with paralysis using an intracortical brain-computer interface. eLife **6**, e18554 (2017)

Sanford, K.: Will This "Neural Lace" Brain Implant Help Us Compete with AI? Nautilus 04 April 2018. http://nautil.us/blog/-will-this-neural-lace-brain-implant-help-us-compete-with-ai. Accessed 11 Dec 2020

Sarayi, S., Mamat, M.Z.B., Hasbullah, M.B.: The human nature: Divinity of man in IBN Arabi's and Shankara's perspectives. Int. J. Innov. Creativity Chang. **8**(5), (2019)

Sarocco, R.: Augmented machines and augmented humans converging on Transhumanism. In: IEEE Digital Reality Initiative (June 2019)

Thorndike, E.L.: The Fundamentals of Learning. Teachers Collage, Columbia University, New York (1932)

Weisberg, S.M., Newcombe, N.S., Chatterjee, A.: Everyday taxi drivers: do better navigators have larger hippocampi? Cortex **115**, 280–293 (2019)

White, B.P., Becker-Blease, K.A., Grace-Bishop, K.: Stimulant medication use in college students: comparison of appropriate users, misusers, and nonusers. J. Am. Coll. Health **54**(5), 261–268 (2006)

Arnsten, A.F.T.: Stimulants: Therapeutic actions in ADHD. Neuropsychopharmacol. **31**(11), 2376–2383 (2006)

Chapter 5
Neo-Luddism

Technology in general has been criticised by many philosophers, intellectuals and movements. Some of the arguments are spiritual in nature, others are based on a romantic notion of the past or the risks of climate change. With the arrival of artificial intelligence, human cognition is challenged as the only known form of intellectual power. With the arrival of a superintelligent AI, human cognition will be surpassed. Artificial intelligence invites its use for almost any human activity. However, intuitively or consciously, many people choose to live without modern technology or at least certain aspects of high tech. How does the introduction of advanced forms of artificial intelligence affect those who choose to live with it and those who choose to live without it?

5.1 Introduction

Maybe some of the developments and their psychological consequences discussed in the previous pages are not quite to your liking? The expectation that neural laces and mobile phones are used for telepathy may not appeal to everybody, in particular if the technology is in very widespread use. The prediction that any advanced form of artificial intelligent should constantly explain the world to humans who are lacking artificially enhanced intelligence may also cause for pause. The anticipated breakdown of the human motivational system is certainly another reason for reflection.

Maybe your instinctive reaction is that it is just about enough now, as far as technological development is concerned. Or that humanity should not embrace a future with an advanced artificial intelligence without very careful consideration of all possible consequences. Maybe your personal decision is not to participate in one or more of these developments and seek a quieter life?

J. Diederich, *The Psychology of Artificial Superintelligence*,
Cognitive Systems Monographs 42,
https://doi.org/10.1007/978-3-030-71842-8_5

You may be a Neo-Luddite. According to the Second Luddite Congress in 1996, Neo-Luddism is a leaderless movement of passive resistance to consumerism and the increasingly bizarre and frightening technologies of the computer age. Neo-Luddism describes an opposition to many forms of technology and in particular information technology.

Luddism was a social movement in England that emerged in the eighteenth century and was characterised by an opposition to industrialisation. The Luddites destroyed machines as symbols of the threat of unemployment to skilled craftsmen at the dawn of the industrial age. The movement used violence and intimidation as forms of resistance against industrial production by unskilled workers who were paid low wages, suffered unbearable work and living conditions and took the jobs of skilled craftsmen who were traditionally organised in guilds (The Luddites at 200).

It is not surprising that the British government outlawed violence against machines at the beginning of the eighteenth century. The penalties for this form of resistance were very severe, in many cases penal transportation to colonies such as Australia. Despite the threat of these penalties, there were Luddite riots in England between 1811 and 1813, at which point the movement was already organised and capable of exercising considerable pressure on government and factory owners. Waves of violence followed by waves of repression including the use of the military to suppress riots. The British economy after the Napoleonic Wars was in dire shape and the last peak of violence by Luddites occurred around 1830.

It has frequently been argued that the Luddite rebellion was against a transformation of work conditions and procedures; it was not directed against machine as such. Violence against humans attracted the most severe penalties and since machines for industrial production were seen not only as a cause but also as symbols of change, the violence was mostly directed against machines that assist production. While in the public's mind Luddites are associated with violence, this is clearly not the case for Neo-Luddites in the twenty-first century.

Neo-Luddites refrain from using certain forms of technology for very personal, religious or philosophical reasons. At the time of the writing of this book, a young lady from Sweden sailed over the Atlantic to protest climate change and the impact of commercial air traffic; a trip that attracted wide-spread public attention. In many cases, a person may not even consciously reflect on why a certain technology is not being used. It may be nothing more than a decision not to get a new smartphone every year. In any case, this new development of Neo-Luddites is most certainly non-violent. As a matter of fact, non-violence is the hallmark of this leaderless movement. A good example are the Amish people in North America. Motivated by anabaptist religious convictions, Amish communities are found in many American states as well as provinces in Canada. This community is known for its pacifist attitudes and almost seamless integration into the fabric of American society.

5.2 Neo-Luddites

Neo-Luddism covers the entire political spectrum. It includes environmental groups who reject certain technologies in order to reduce humanity's carbon footprint along with back-to-nature communities that are motivated by a form of nineteenth century romanticism. It includes anarchist artists who reject civilisation, or at least Western civilisation, and religious communities who are part of a new wave of monasticism. Nevertheless, these groups share a number of relevant features. They are non-violent, they tend to live on the land and the numbers are increasing rapidly. The North American Amish population grew by an estimated 172,780 since 2000, increasing from approximately 177,885 in 2000 to 350,665 in 2020—an increase of 97 percent (Amish Studies, The Young Center). Young people as a driving force is another characteristic; these are most certainly not retirement communities.

Kryszczuk and Wenzel (2017, p. 45) developed a taxonomy of Neo-Luddist movements: Economic Neo-Luddites (1) act on anti-libertarian instincts and advocate re-agrarisation or deindustrialisation to prevent unemployment and social decline. Romantic-pastoral Neo-Luddism (2) is associated with environmental movements, the questioning of modern mass production and the promotion of vegetarianism and escapism. Furthermore, Kryszczuk and Wenzel (2017, p. 45) identify (3) spiritual-ideological Neo-Luddism. This movement is characterized by scepticism toward mainstream science and technology. A separate strand is Anarchic Neo-Luddism (4) which draws inspiration from the violence used by original Luddites (Kryszczuk and Wenzel 2017, p. 45).

Why would anybody choose to live without the benefits of modern technology? Barber (2008, in Kryszczuk and Wenzel 2017, pp. 52–53) offers one view:

> Consumerism urges us to … step into the new world of electronic toys, games and gadgets that make up the modern digital adult playground; the market apparently recognized that people do not have to grow up anymore. Instead of getting schools to help children grow out of toys, we bring toys to schools - video games and computers as learning aids, as well as commercially sponsored television in classrooms. … Traditional teaching methods cannot compete with the charm of the world of commercial games …

Mark Boyle would be an example of the second category of the Kryszczuk and Wenzel (2017) taxonomy related to environmental concerns, a critique of mass production and escapism (Boyle 2019). Mark Boyle is an author and columnist of the Guardian newspaper. Mark experimented with several forms of off-the-grid living over the last years, including living without money for a period of three years. His publishing activities allowed Mark to purchase a cabin and some acreage in Ireland in order to experiment with living without technology. Mark is not just avoiding information technology, he avoids all forms of modern aids including washing machines and other devices. In his autoethnographic book (*The Way Home*, Boyle 2019), he reports on the benefits and challenges of his new life. This includes writing articles and books by use of pen and paper as well as the long process of heating water to have a bath. He emphasises that is not just living without technology that is the key to "coming home", it is also the avoidance of clocks. Without a "time giver" such as a clock, life

Fig. 5.1 My friend Kieth

returns to the rhythm of daylight and the seasons. However, the benefits of natural living come at a cost: everything takes much longer (Fig. 5.1).

Recently, psychologists looked at the benefits and disadvantages of living in the city as opposed to a more rural lifestyle (Snell and Bhullat 2019). Humans have lived in cities for about 5000 years since the emergence of agriculture. Clearly, the high density living that characterises current mega-cities is a more recent development. The proportion of people living in cities has increased dramatically over the past 200 years. Just 3% of the worldwide population lived in cities in 1800 while currently 55% of all individuals live in an urban environment. In the year 2050, this number is projected to increase to 68% (Snell and Bhullat 2019, p. 9). Even in a country like Australia with a significant land mass, two-thirds of all inhabitants live in the major cities.

The move to the city is driven by the availability of services and jobs as well as cultural and social opportunities. The availability of public transport in particular results in an urban lifestyle that is completely different from living on the land. However, psychologically urban living comes at a cost. Residents of urban areas have to cope with the effects of overcrowding, high noise levels, dense traffic and frequently pollution (Snell and Bhullat 2019). In addition, living in a city is expensive and there is the risk of exposure to crime and the resulting trauma.

On the other side, regional or rural living risks social isolation and loneliness. The relative scarcity of specialist medical services as well as long commute times can also be seen as psychological risk factors. Without careful consideration of the costs and environmental impact, rural living is less energy efficient then an urban lifestyle.

Historically, a famous Neo-Luddite was the German philosopher Martin Heidegger who enjoyed spending time in a small cabin in the Black Forest mountains of Germany. Heidegger said at some point that he prefers the company of farmers to those of professors. While the hut is still private property and belongs to the family, many pictures are available, in addition to a book published by MIT Press. The images show very simple living condition even though electricity is available. This simple rural cabin is so famous that the Black Forest tourism authority organised a walking trail around the property which offers views from different perspectives. The overview of Sharr's book from 2006 probably describes the fascination surrounding this simple property best:

> There are many ways to interpret Heidegger's hut—as the site of heroic confrontation between philosopher and existence; as the petit bourgeois escape of a misguided romantic; as a place overshadowed by fascism; or as an entirely unremarkable little building.

5.3 "Only a God Can Save Us": Heidegger's Spiegel Interview in 1966

Martin Heidegger gave a famous interview with the German magazine "Der Spiegel" on 23 September 1966. Heidegger insisted at the time that this interview can only be released after his death and the German magazine published it five days after his death on 31 May 1976. In this interview, Heidegger responded in detail to questions about his role during national socialism in Germany and in particular his time as vice-chancellor of the University of Freiburg.

Heidegger's role during the period of national socialism in Germany is controversial to this day. As a matter of fact, the publication of the "Black Notebooks" inflamed the discussion in recent years. Heidegger joined the Nazi Party in 1933, the year the national socialists came to power in Germany and he became vice chancellor of the University of Freiburg the same year. He left his post as vice chancellor in 1934 and his performance as a university administrator is generally regarded as a failure. Nazi imagery was used during public appearances while he was vice chancellor. While he became increasingly distant from everything associated with national socialism after his time as vice chancellor, he did not reject national socialism with absolute clarity after the Second World War (Wheeler 2018). It is this lack of clarity that has been criticised most.

One of the most important allegations against Heidegger is that he dreamed for some time during the early 1930s about being the leader of the leader ("der Führer des Führers"). The illusion was to become some form of intellectual guide to Adolf Hitler. And with Hitler to guide people back to a more original form of living. He soon realised that national socialism had nothing to do with the romantic nationalism of the nineteenth century; this was purely about race and territory as well as an acceleration of technological development for the purpose of war and the suppression of those who were deemed inferior. The Nazis developed technology not for the experience of being in the world, but for destruction and death.

Heidegger's main work *Being and Time* ("Sein und Zeit") was published before the period of national socialism in Germany from 1933 to 1945. The later part of his work is often described as *The Turn* ("Die Kehre") and was written after the Second World War. The most disturbing aspect of the publication of the Black Notebooks is that Heidegger held primitive anti-semitic views even after World War II (Rothman 2014). While Heidegger's association with the ideology of national socialism is simply unacceptable, few would go as far as to completely ignore his contributions as an existentialist philosopher in the twentieth century. The name Heidegger is also linked with a thorough critique of technological development and this is important in this context.

The second part of the *Spiegel* interview from 1966 moves away from Heidegger's activities during the period of national socialism and touches on the role of technology in today's society. The headline of *Der Spiegel* for the interview is "Only a God can save us."[1] Heidegger may not necessarily refer to God as a universal creator; he uses a secularised notion of the sacred (Wheeler 2018). Or to put it another way, technology cannot save us from technology (Trawny P, personal communication 2020).

In the interview, Heidegger expressed his shock of seeing the first pictures of Earth from the moon. He then talks about the very essence of technology and human control.[2] According to Heidegger, modern technology is not a tool and it has nothing to do with "tools". He states that there has not been any real reflection on technology ("Technik") since the view dominates that technology is something that humans can control. Technology by its very nature, according to Heidegger, cannot be controlled by humans.

Following Heidegger, humans are presented with, occupied with and challenged by a force that is present in the very nature of technology. The traditional approach of metaphysics which was finalised with Nietzsche offers no way to approach the essence of technology in this new period of history. *Humans are not overwhelmed by technology, they simply have no way forward that corresponds to the nature of technology.* It is the reliable functioning of technology that cuts the human link to Earth.

Continuing with Heidegger's argument, everything important and great was possible because humans had a home and were rooted in history. Yet, traditional

---

[1]It is interesting to note that at the issue of the magazine *Der Spiegel* from 31 May 1976 includes advertisements as well. Page 201 includes an ad for a prescription free medication to enhance the sexual performance in human males. An advertisement for similar product is on page 217. These ads are oddly out of place from the perspective of a reader more than fifty years later. Today, they can be seen as advertisements for transhumanist enhancements. But maybe this is just what the marketing department of *Der Spiegel* put on these pages.

[2]The interview was conducted in German language, and with everything Heidegger, language is important. Some of the English translations are decades old and not by authors with a background in technology.

philosophy is now replaced by cybernetics. At this point, Heidegger introduces the notion of God. Only a god can save us, not philosophy. Our only chance is in thought and poetry to be prepared for the appearance of God or to disappear in the absence of God. To be prepared and open for the arrival or the absence of God.

Feenberg (2010) writes

> Heidegger's critique of technology is ontological, not sociological. This ontology is so contrary to common sense that it is very difficult to understand. We tend to think that reality is "out there" while our consciousness is an inner domain that gains access to things through the senses. Heidegger rejects this model.

Feenberg (2010) continues

> As Heidegger explains it, our most basic relation to reality is not perception as we usually understand it. That is a theoretical construction. Abstracting from our actual experience, we … encounter our world not through causal interaction between nature and the senses, but rather through action directed at meaningful objects. These primordial encounters later become objects of reflection …

Heidegger uses the German word "Ge-stell" to describe the essence of technology (most English translaters use the word Enframing).[3] Technology is not, or not just, a tool or a means-to-an-end, *it is a mode or form of human existence.* Heidegger (2000, p. 25) writes

> Ge-stell "ist nichts Technisches, nichts Maschinenartiges. Es ist die Weise, nach der sich das Wirkliche als Bestand entbirgt. Wiederum fragen wir: geschieht dieses Entbergen irgendwo jenseits alles menschlichen Tuns? Nein. Aber es geschieht auch nicht nur im Menschen und nicht maßgebend durch ihn" (Ge-stell is not technological, nothing like machines. It is the way reality offers itself as real. Again we ask: is this "Entbergen" (Revealing) somehow beyond human activity? No. However, it does not just happen in humans and not significantly through human action).

"Entbergung" through technology is "Wahr-heit" (truth). Enframing refers to the way technology presents itself as part of the world, how it appears to us. But as Heidegger suggests above, Ge-stell goes way beyond the simple appearance of technology. There is an active aspect to it *since technology wants to be used.* It does not only suggest a certain usage but a mode of existence and an important aspect is that once technology is introduced into the world, it is nearly impossible to go back.

A discussion of technological use invites a description of the term affordance. Affordance is an aspect of an object that suggests or requires a certain usage. The affordance of a door handle is to push down on it, the affordance of a doorknob is to turn it. Artificial objects that we call technology have affordances. A mobile device such as a tablet computer has affordances such as pressing and swiping.

---

[3] It may be difficult for English readers to appreciate that Heidegger almost created his own language by use of German expressions. This language is very difficult to understand even for native speakers of the German language. And it makes translations into English particularly problematic. Many translations, and in particular those that touch on Heidegger's concerns about the role of technology, are hopelessly out of date. Hence, I am providing my own translations here. These translations may be as problematic as previous attempts, and therefore, the text of the German language original appears first followed by my translations.

Yet, we are still missing out on the essence of technology. Many three or four year old children can use mobile devices such as tablets without any problem. They know about pressing a button and swiping but they do not know how the technology works. In the world of the developing child, there is a flat thing that shows movies as well as images and it allows you to talk to your parents even when they are away. Once the object of a tablet computer has been introduced to a child, for instance, there is no way to go back. There is no way to go back to the cinema of one-hundred years ago and there is no way of going back to a time without movies. Mobile devices allow and suggest watching movies, they are a significant part of being in the world, *even the world of those who choose to live without mobile devices and movies. Even if you decide to live without a smartphone, the solicitation remains.*

To the transhumanist, the human experience is painfully insufficient, so the observation that technology, once introduced, does not easily allow the reversal of its introduction, is a non-issue. Yet for Heidegger, modern technology blocks access to truth. Technological progress results in forgetting what it means to be and any form of non-technological way of perceiving and knowing (Fuchs 2019). Furthermore, there may be a loss of experience and knowing through the arts and artistic activity. Technology is a problem if it is the only way of knowing and perceiving. In this sense, Heidegger calls technology a perfect form of metaphysics and a perfect loss of knowing what it means to be (Fuchs 2019). Hence, technology represents the ultimate danger.

> Das Geschick der Entbergung ist in sich nicht irgendeine, sondern die Gefahr. (The fate of the "Entbergung" is not arbitrary, it is danger as such; Heidegger 2000, p. 27).

Technology is the ultimate risk. However, where there is danger there is also rescue (Heidegger 2000, p. 29):

> Die Herrschaft des Ge-stells droht mit der Möglichkeit, daß dem Menschen versagt sein könnte, in ein ursprünglicheres Entbergen einzukehren und so den Zuspruch einer anfänglicheren Wahrheit zu erfahren. So ist denn, wo das Ge-stell herrscht, im höchsten Sinne Gefahr (The rule of the Ge-stell threatens the possibility that humans cannot return to an original form of Entbergen and hence an original way to experience truth. Ultimate danger exists where technology rules).

Heidegger quotes the German poet Hölderlin: "Wo aber Gefahr ist, wächst Das Rettende auch" (Rescue grows in the presence of danger; Heidegger 2000, p. 29).

A turn ("Die Kehre") is possible. Access to the truth that is blocked by the Ge-stell is restored, motivated by an awareness of danger. Since all these considerations are quite abstract, the Rhine river is used as an example. This river has a special place in German mythology, history and literature. There are so many references to the beauty of the river that certain sections became symbols of romanticism (e.g. Lorelei). Imagine now a hydroelectric dam would be built on the river Rhine. Immediately, this dam changes the meaning of the river; it becomes an energy resource. The notion of a power source literally blocks the beauty of the river from perception and experience. Let's further assume that the decision would be made to remove the hydroelectric dam, since other power sources are readily available. The removal of the dam would then enable the original romantic experience of the river, but only after some time.

But is a true return to the previous state really possible? Let's look at another example. Decades after the digital revolution in musical recording and playback, the analogue vinyl format makes a comeback. There are enthusiasts everywhere who believe in the better sound quality of analogue records and as a result, there is a new market for record players and analogue recordings. Essentially, a subculture has emerged of true believers who like to listen to music old style. This is different to the situation in the middle of last century. People bought analogue records because this was the only way to go. It did not require any reflection or identification with a subculture. If you wanted to listen to music, you bought a record. In a sense, the technology was ready-to-hand. Now it requires an effort to get the equipment as well as the records, and it even may require some form of defensive argument since most people believe that digital music is superior. It can be argued that no real return is possible.

5.3.1 *Artificial Intelligence Is Not (Just) a Tool*

A situation has a way of drawing one into it (Dreyfus 2007, p. 252). Tools have affordances or solicitations. A hammer has the solicitation to grab it. A smartphone has too many solicitations to list here. There is a "ready-to-hand" or "pressing-into-possibilities" that occurs (Dreyfus 2007, p. 252):

> For Heidegger the ready-to-hand is not a fixed function, encountered in a predefined type of situation that triggers a predetermined response that either succeeds or fails. Rather, as we have begun to see and will soon see further, readiness-to-hand is experienced as a solicitation that calls forth a flexible response to the significance of the current situation.

We cannot know the ready-to-hand aspects of an artificial superintelligence and we cannot know into what form of possibilities we are pressed into. We know from today's artificial intelligence that it can be used for manufacturing, reading or listening, for search and marketing, and it is possible to have sex with robots that have AI components. Humans are and will be pressed into many things and this includes using the off-switch if one is available. However, a human being who approaches an AI system to turn it off has been pressed into many other possibilities already even if this did not result in behaviour. So it is the *universal solicitation of AI that is the problem, or more precisely the totality of all solicitations.* Another way to put it is that the damage is being done by this thing being in the world, not by what it does or what it can do. Turning an advanced artificial intelligence off will shut down functions but it will not remove it from the world.

There are a few takeaway messages from this discussion surrounding Heidegger and his thoughts. Any understanding of artificial intelligence as purely a tool, instrument or machine falls short. Artificial intelligence changes the way people experience the world, but it also changes the world view of individuals who choose not to use the technology. In the example of the Rhine river above, it is not necessary to have seen the hydroelectric dam, simply knowing that it exists will trigger the shift from romantic beauty to energy resource. Neo-Luddites may have an intuitive or explicit understanding of the risk that AI poses, yet they are nevertheless affected by it since total isolation from the technological world is not possible.

This psychological impact of technology is crucially important. While the super-intelligent robotic butler of the future may save time and effort, it may also block the experience of an original form of reality. Clearly, the users of AI in the future will have to make very personal and important decisions.

In summary, if technology is not a tool ("Werkzeug") then AI is most certainly neither. If humans cannot control technology in general, then they cannot control AI. The search for the "off switch" is certainly falling short since it does not address the nature of technology as such. Whoever is reaching for the off-switch has been significantly exposed to the technology and its risks a long time before the thought emerged to "switch off".

5.4 A Thought Experiment

As part of a thought experiment, imagine you would live one day without technology. Mind you this is a thought experiment only, since living without technological help may be dangerous.

Let's think through how it would be to live without technology for a single day. First, we want to make sure that we are not relying on technology without knowing it. It is just too easy to grab a mobile device for navigation and other simple tasks. So to make sure that there is no advanced technology whatsoever, we turn off the electricity in the home and make sure we have no battery power as well.

As part of our thought experiment (please do not do this if you rely on technology for health and safety reasons), we have now disabled most if not all of the electronic technology and we are relying on very simple mechanical tools. Immediately, it gets very busy. If it is cold outside, we have to collect wood for heating purposes. If it is too warm outside, cooling needs to organised in other ways, e.g. withdrawing to rooms or outside areas that are tolerable. If it is cold and a fire has been lit, there is certainly work involved to keep it going, so the process requires attention every 30 or 60 min or so.

Next, food has to be purchased and prepared which may include the fireplace as well. If only one vessel is available for cooking then obviously the entire process takes longer. Once food has been prepared and consumed, cleaning of dishes and clothing maybe a next step. Naturally, the laundry takes much longer without any help from technology.

At this point the realisation may hit that this kind of lifestyle is hardly compatible with an 8 to 10 hour workday. If travel to and from work is involved, plus a stressful job, then all the extra duties that appear on the horizon without the support of modern technology makes the endeavour extremely hard. So while we have turned off electricity to avoid the use of technology for one day, we will have by necessity completely reorganised the home and work lives at the same time.

As evening approaches, there's obviously no TV but the reading of books is possible by use of oil lamps. News junkies may have purchased the print copy of a newspaper. If you have been troubled by wars far away, then maybe less so during the day of our thought experiment. If you have been troubled by the constant flow of opinions and emotions on social media, then this may be less so during the time of our thought experiment. Why worry about something that you cannot read and hence cannot know? But there is something else that is even more important psychologically.

All the additional labour that is required forces the participants in the thought experiment to focus on the immediate environment and necessary tasks. The thoughts are occupied by the task of keeping the fire going as well as preparing food. All these aspects of the immediate environment occupy the mind, and not so much events, actions and opinions far away of which we can have no knowledge during this day.

So, what started as an experiment of avoiding technology for one day became an *exercise of mindfulness*, the psychological technique that assists in the recovery from depression, anxiety and panic disorder. Mindfulness is a

> capacity of attention and awareness oriented to the present moment that varies in degree within and between individuals, and can be assessed empirically and independent of religious, spiritual, or cultural beliefs (Black 2011).

Mindfulness is a method to cope with tension which advances well-being. It is the cognitive process of focusing on the present moment without judgement. Focused action requires mindfulness.

5.4.1 Mindfulness in Psychology

Mindfulness is an intentional, open, receptive and non-judgemental form of attention, a focus on what is happening in the present moment. Mindfulness avoids habitual responding and scattered attention (Black 2011). In contrast, *mindlessness* occurs if there is a preoccupation with past experiences or future events and hence a limited awareness of the current environment. Individuals who suffer from extensive worrying and rumination detach themselves from the here and now, take their minds to the future and play through all possible scenarios of the things that can happen. If these scenarios are negative, then this represents a form of unhelpful thinking style which is part of depression according to cognitive theories of the disorder.

Brown and Ryan (2003, p. 823) write

> Mindfulness captures a quality of consciousness that is characterized by clarity and vividness of current experience and functioning and thus stands in contrast to the mindless, less "awake" states of habitual or automatic functioning that may be chronic for many individuals. Mindfulness may be important in disengaging individuals from automatic thoughts, habits, and unhealthy behaviour patterns.

Current technology facilitates mindlessness. Smartphones offer many opportunities for distractions, a multitude of visual cues and opportunities to jump from one topic to the next without any logical coherence. Online advertisements in their current form most certainly encourage mindlessness since content is offered based on the strategies of advertisers and not the needs of a user. Most mindfulness techniques in psychology quite literally require users to switch off technology.

Maybe techniques such as mindfulness would not be necessary without the effects of technology? As much as technology supports or replaces human action, it also detaches the mind from reality (i.e. the current situation). Since there is reduced awareness of the surrounding environment, simply because it is not required, the individual has cognitive resources detached from reality and can turn their thoughts inside. Rumination is possible because the person is not occupied with all the immediate tasks and actions. Being present in the immediate environment and doing things in the "here and now" reduces the opportunity for rumination and all the unhelpful thinking styles that many individuals experience in anxiety and depression.

5.4.2 Husserl's Phenomenological Reduction

While mindfulness as it is taught today by clinical psychologists is closely related to Asian philosophy, it has obvious links to Western philosophy and in particular to Edmund Husserl's phenomenology. Husserl, who heavily influenced Martin Heidegger, introduced the term of bracketing ("Einklammerung") as a form of perception that is free of judgement and based on the suspension of believe in the objective nature of the word. The goal of the process is to arrive at experience and perception as such, freed from prejudice and symbolic meaning to the largest extend possible.

Husserl called this mode of processing the phenomenological attitude (Brown and Cordon 2009, p. 62)

> in which our attention is turned toward reality simply as it appears or is given to us, that is, simply as a flow of phenomena or appearances.

Husserl called the process of doing so "phenomenological reduction". This does not mean a replacement or an elimination of cognitive operations but rather a "stepping back" from our usual way of processing in order to receive experience as it manifests itself to us. In this way, everything—sense impressions, feelings, images, and thoughts—remain but are perceived in a different way, that is, strictly as they appear (Thompson and Zahavi 2007). In this stepping out of the natural attitude, through a "suspension" or "bracketing" of our habitually conceptual mode of processing, the mind discloses how reality is "constituted" in the present moment and within the structure of our conscious minds.

5.4.3 *Heideggerian AI*

With Heidegger's general attitude towards technology, it may come as a surprise that there is something like a "Heideggerian artificial intelligence". The critique of symbolic artificial intelligence, from a Heideggerian perspective, is closely associated with Hubert Dreyfus. To be truly in the world, it is argued, intelligence has to be embodied. There needs to be the experience of gravity, moving around, learning to grasp objects and interacting with other embodied minds. From this perspective, a cognitive AI system must be a robot.

Dreyfus outlines a form of AI that follows Heidegger (Dreyfus 2007, p. 249):

> Heideggerian AI that might be thought of as articulating a new paradigm for the field: Rodney Brooks' behaviorist approach at MIT, Phil Agre's pragmatist model, and Walter Freeman's dynamic neural model. All three approaches accept Heidegger's critique of Cartesian internalist representationalism, and, instead, embrace John Haugeland's slogan that cognition is "embedded and embodied".

Behavioural robotics aims at designing machines that cope with situations without drawing on explicit, internal representations. If an artificial superintelligence will ever be based on the principles of behavioural robotics remains to be seen. Without explicit models of the external world and without the ability to reason about actions and their consequences, behavioural robotics is limited.

5.5 Resistance to Artificial Intelligence

The previous sections were about partial or complete resistance to modern technology. What about a resistance to artificial intelligence at this point in time? There are several areas where concerns have been expressed internationally about AI developments, including: (1) lethal autonomous weapons that include artificial intelligence, a point that will be explored more deeply in the chapter on safety and military artificial intelligence, (2) bot technology for misinformation, (3) AI and the invasion of privacy and (4) medical artificial intelligence. Each of these will be discussed further in the sections below. Please note that resistance to AI is different to the situation where human pose a risk to AI.

5.5.1 *Lethal Autonomous Weapons*

The concern about lethal autonomous weapons is really focused on AI killing machines without a human in the loop. In this scenario, there is no human decision-making or human review of who should be killed and when. That is, the decision about the use of deadly force against an enemy combatant is completely delegated to a machine. Obviously, various forms of human oversight are possible, however, this does not eliminate the concerns that many researchers and practitioners express.

Assume a group of autonomous robots with which lethal weapons that operate as a swarm in an urban environment. These robots can search streets, buildings and eliminate targets without risks to soldiers of the home team. This is a cheap and attractive solution. No equipment or resources are required to protect human soldiers fighting on the same side as the autonomous robots. This simplifies logistics, and at the same time, lowers the threshold for the use of violence: if your own team is not being hurt and risks posed by enemy combatants can be eliminated, then why not simply go ahead? Obviously, the enemy combatant is a human being and if all sides use this strategy, then we are looking at an all-around massacre.

In an interview with *The Australian* newspaper on the 16th of June 2020, the Professor of Artificial Intelligence at the University of New South Wales Toby Walsh stated that autonomous drones that independently decide if and when to kill are being deployed already (Griffith 2020). The interview reported on a keynote address given at the 2020 International Conference on Robotics and Automation. Toby Walsh refers to the use of the drones along the border between Syria and Turkey. He claims countries such as Pakistan are using the machines as well.

Toby Walsh said "the drones are fully autonomous and use AI and other face recognition software to identify, track and kill people on the ground". He claimed the drones have been developed by Turkey after a student who previously worked at MIT developed the technology in a very brief period of time. In the same article it was reported that Russia is planning a global network of autonomous underwater, surface and air drones and that countries such as China are already selling autonomous killer drones. The article explores the risk of the technology being used by terrorists against a civilian population. The implications of this type of technology will be explored in more detail in the chapter on military AI and safety.

There is nearly complete, international consensus among AI researchers that the development of autonomous, lethal AI systems should be outlawed and that these systems should not be deployed. Heidegger's concerns about the ready-to-hand character of technology comes to mind again. Autonomous, weaponised drones are a cheap ready-to-hand solution to a problem that has cost the lives of many soldiers in the past: search and destroy urban warfare.

The naive response to the concerns outlined above is to remove all human combatants *on all sides* from the battlefield. If there are no humans, nobody gets hurt. This would lower the threshold for using AI guided violence even further, not just in the case of war, but in everyday life as well, since the technology is readily available. Besides, unlimited robot wars are clearly not an option in a world inhabited by humans. To return to the psychological argument, is it reasonable to be afraid of a world with AI warfare? There are too many science fiction movies to list here that play out the scenarios mentioned above. It is plausible to say that these visions cause fear.[4]

[4]Netflix' Black Mirror, Season 4 "Metalhead" certainly scares me.

5.5.2 *Monitoring Online Conversations*

In the follow-up to the US elections in 2016, concerns were expressed about the role of conversational bots on social media. Bots can initiate conversations, respond to messages as well as liking or disliking certain statements. The concern was (and still is) that the technology is being used to shape conversations and as a result, the outcome of elections.

An additional issue is the nature of online conversations on social media, and in particular, the use of abusive and threatening language. Consequently, technology has been developed to monitor the "toxicity" of online conversations. Jigsaw, a Google based technology incubator, developed the Perspective API to monitor online trolling and hate speech. Perspective is described as follows (Jigsaw 2018):

> Perspective is an API that makes it easier to host better conversations. The API uses machine learning models to score the perceived impact a comment might have on a conversation. Developers and publishers can use this score to give realtime feedback to commenters or help moderators do their job, or allow readers to more easily find relevant information …

The Perspective app is based on machine learning and requires labelled data for training purposes. The collection and labelling processes are obviously crucial and are described by the creators of the Perspective API as follows (Jigsaw 2018):

> Our data collection methodology begins by asking crowd-workers to measure the overall level of "healthiness" in a comment. The goal is not to be overly prescriptive … We want to allow for many different types of conversation as long as they meet a basic standard of healthy engagement.
>
> We define five 'sub-attributes' … (1) hostile, antagonistic, insulting, provocative or trolling; (2) dismissive, (3) condescending or patronising, (4) sarcastic, and (5) unfair generalisations … We believe these attributes enable us to capture most of the more subtle kinds of toxic language discussed above.

The obvious concern is that a group of people in a technology project define what a healthy conversation is for the rest of the planet. To what extent is this particular group aware of their own biases? And if this particular group determines what a healthy conversation is for the rest of society, who is monitoring or controlling this particular group? It would be easy to point to areas where unhealthy or toxic conversations can occur and this includes psychological therapy and assessment. As outlined in the chapter on digital clones, speech and language are very important for psychological assessment and provide crucial information for clinicians.

Woolley (2020) criticises the use of applications such as Perspective. There is obviously the problem of false positives, i.e. statements or remarks that have been incorrectly identified as hateful or toxic. The usefulness of the application depends on the data labelling process and this is a very culturally sensitive.

5.5.3 AI and Privacy

#MeTooBots are currently being developed to identify digital bullying and sexual harassment by scanning emails and other forms of communication. The tools are currently offered by the Chicago-based AI company NextLP and are designed to monitor communications between employees in order to flag messages that indicate sexual harassment or bullying (Curtis 2020). In early 2020, it has been reported that more than 50 corporations around the world are using the technology, including law firms.

The use of such technology has been criticised for a number of reasons. Allegedly, #MeTooBots assume that digital misbehaviour and sexual harassment are the default states in workplace environments. Hence, all communication needs to be monitored to eliminate harassment. The problem arises that simple pattern recognition algorithms are not sensitive to the context of communication and therefore very limited in their application. More importantly, it is possible to manipulate machine learning algorithms by providing fake or misleading training data. A "denial of service" type attack on a machine learning system trained to identify digital harassment would swamp the algorithm with data indicating harassment or not. This would have a significant impact on the performance of the system and therefore, would have to be monitored very closely.

Another critique is that digital harassment detection algorithms creates an atmosphere of mistrust in the workplace environment. It can result in suspicion and self-censorship. Hence, the social consequences may outweigh the benefit of pattern recognition algorithms for digital misbehaviour.

5.5.4 Resistance to Medical AI

Resistance to AI projects or the adoption of artificial intelligence within organisations is commonly attributed to cultural issues or simply a lack of information or education. If employees and managers do not recognise the need for artificial intelligence, the solution is that they need to be educated. Frequently, this is being done by providing use-cases from other parts of the industry to point to the need and potential success of AI projects.

However, there are deeper psychological issues. There is a tendency, for instance, not to trust the forecast of statistical models even if they are superior to human decision-making (Meehl 1954, in Longoni et al. 2019). In other words, even though statistical models outperform human intuition when making predictions, individuals still prefer human expertise when making their own decisions. Currently, there is a reluctance to trust the decision of an artificial intelligence system in the medical field even if the system is not statistical in nature and has proven superior to medical experts (Longoni et al. 2019).

Yet, artificial intelligence is revolutionising the field of healthcare (Ahuja 2019, Panch et al. 2019). AI systems have proven effective in the diagnosis of heart disease, in providing medical advice as part of National Health Services as well as the detection of skin cancer (Longoni et al. 2019). Even though all these systems perform with expert level accuracy, there is still a reluctance to accept decision making of an AI system. Longoni et al. (2019, p. 631) write

> We propose that consumers may be more reluctant to utilize healthcare delivered by AI providers than healthcare delivered by comparable human providers, even when they are explicitly informed about the performance of the providers. We further propose that this occurs because the prospect of being cared for by an automated provider evokes a concern that one's unique characteristics, circumstances, and symptoms will be neglected. We refer to this concern as uniqueness neglect. We argue that uniqueness neglect emerges out of a mismatch between two fundamental beliefs. Whereas consumers view themselves as unique and different from others …, consumers view machines as capable of operating only in a standardized and rote manner that treats every case the same way.

It is easy for a clinical psychologist to relate to this phenomenon of *uniqueness neglect*. Frequently, clients present with concerns in daily clinical practice that they describe as completely out of the ordinary. Clients may even apologise for presenting with concerns because of their perceived uniqueness and the view that a certain presentation separates the individual from the rest of society. Once the clinician has conducted the assessment, it may turn out that the core of the concern is related to depression and anxiety, i.e. psychological problems that are very frequent.

If it is part of the cognitive self model that individuals view their own behaviour and thought processes as unique, how can a statistical model or AI system then accommodate for this uniqueness concern?

The answer is of course that people who live in the same society are exposed to the same stressors and risk factors that contribute to mental health problems. People share cultural norms and expectations as well as similar family and work environments. As a result, individuals tend to develop similar mental health problems. The cognitive dysfunctions and unhelpful thinking styles that can cause depression are a very good example: many people catastrophise or personalise their problems or use emotional reasoning. As a result, there is an overlap in psychiatric symptoms and true uniqueness is rare.

However, if the uniqueness neglect is an obstacle to the adoption of AI systems in the medical sector, then the question is how AI integrated into healthcare can contribute to a patient's satisfaction.

Outside of the area of medical AI, Luo et al. (2019) studied the effectiveness of artificial and human conversational agents in making outgoing calls for marketing and customer management. In some countries or territories (e.g. California), it is now a legal requirement for a chatbot to identify itself as an AI system. Given that human and chatbot interactions can be mixed arbitrarily, it is an ethical and legal requirement to have proper identification at the point of change. In many applications, the chatbot starts a conversation, identifies the goals and objectives of the customer and then hands over to a human operator within the same application. Once human interaction has finished, the AI conversational agent takes over again to address further queries or for quality assurance (an example is the Lifeline bot in Australia). As discussed already in the chapter on digital clones, the change-over from human to AI interaction and vice versa is crucially important.

Google Duplex is a free artificial intelligence service that makes restaurant reservations by use of a human voice. Google Duplex is based on a recurrent neural network and makes outgoing calls first by searching for a list of available restaurants and then by placing reservation with them. In 2019 it was reported that in about 25% of outgoing calls, a human operator was used for part of the conversation (Chen and Metz 2019). While the conversation is very natural including delays and hesitations, there is currently no way for a user to listen to the outgoing call or to receive a written record. Currently, Google Duplex is available in most states of the United States and it is anticipated that the service will become part of Google Assistant.

Luo et al. (2019, p. 913) report on

> …. experimental data on more than 6,200 customers who are randomized to receive highly structured outbound sales calls from chatbots or human workers. Results suggest that undisclosed chatbots are as effective as proficient workers and four times more effective than inexperienced workers in engendering customer purchases. However, a disclosure of chatbot identity before the machine–customer conversation reduces purchase rates by more than 79.7% (Chandler 2019).

Luo et al. (2019, p. 913) find that the chatbot disclosure substantially decreases call length. When customers know the conversational partner is not a human, they *are curt and purchase less* because they perceive the bot as less knowledgeable and less empathetic. The negative disclosure effect seems to be the result of a subjective human bias against machines.

The observations by Luo et al. (2019) appear to be similar to the uniqueness neglect phenomenon in the sense that customers believe that their requests are distinctive to the point that no machine can adequately respond. However, Luo et al. (2019, p. 913) write that the negative impact can be mitigated by a late disclosure timing, that is, telling the customer relatively late in the conversation that responses are generated by a bot.

5.5.5 Determination of Ethnicity, Personality and Attractiveness

Currently a number of free websites allow the upload of personal images and offer a judgement of attractiveness. Services like hotness.ai provide a ranking of physical attractiveness on a scale from 0 to 10. The same company haystack.ai offers an assessment of other facial features including emotion, gender and ethnicity. This service has been adopted by ancestry.com in order to determine the ethnicity of a person in passport-type images. Another system developed by the University of Melbourne called Biometric Mirror offers emotion and kindness classification as well as a judgement of happiness, common sense and responsibility (Bennett 2018). Obviously, the

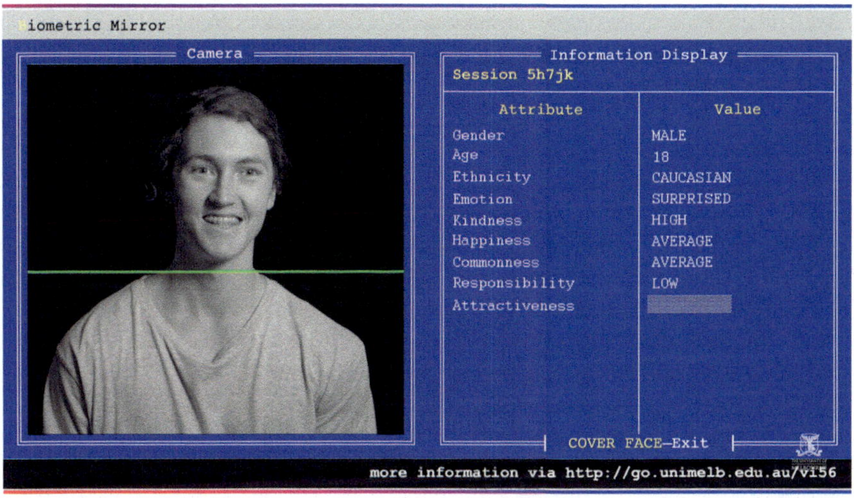

Fig. 5.2 Biometric Mirror uses an open dataset of thousands of facial images and crowdsourced evaluations. Image: Sarah Fisher/University of Melbourne. Lead Researcher: Dr. Niels Wouters

system cannot determine a person's real personality, yet the output would indicate that an attempt is made. The authors admit that this particular application of artificial intelligence is a slippery slope (Gorey 2018, Fig. 5.2).

No studies could be located that demonstrate the impact of this type of technology on people who are psychologically vulnerable, including children and adolescence, individuals with mental health problems and personality disorders. Again, while no studies are available at the time this book is written, the impact of the technology on persons with body dysmorphic disorder or a narcistic personality problem is assumed to be significant.

In a complete surveillance environment, the permanent judgement of personality and features such as attractiveness must have a significant impact on a person's well-being.

5.6 Societal Response to Advanced Technology

There are currently several trends or movements against the development and use of technology. Some are directed against technology in general, some are more specifically directed against forms of artificial intelligence. Some concerns are philosophical in nature, some are about the psychological impact and others are about the security of individuals and society in general. These movements and schools of thought interact. For instance, environmental concerns may result in a rejection of some aspects of globalism including air travel and a promotion of a simpler form of life based on local interactions only. While governments push massive investments into artificial intelligence (both China and Germany serve as examples), there is clearly a desire by some to return to a simpler life.

There appear to be two processes in western society that are currently on a collision course. On the one hand there is a tendency to return to a simpler way of living due to spiritual or environmental reasons, or due to concerns about the ever increasing role of technology. On the other hand, there are billions of dollars that governments and private organisation invest in the advancement of artificial intelligence including military applications. Artificial intelligence has seen waves of funding since its invention in 1958 and there have been several so-called AI winters as well, when funding dropped and disappointment dominated. Currently, we are at the peak of an AI hype mostly due to advancements in machine learning and in particular deep learning.

At the same time, the fear of a global surveillance state and the use of autonomous, lethal weapons rank as primary concerns against both the development and use of artificial intelligence. It is interesting to note that while there is a wide diversity of views in most societies, many people reject the use of killing machines that operate without human input and the invasion of privacy driven by surveillance technology. Nevertheless, there currently is no movement against these applications of artificial intelligence that mirrors the concerns about climate change.

The universal solicitation of AI is always present, and as discussed above, technology can disconnect a person from reality, sometimes in an unhealthy way. From this perspective, a superintelligent robotic butler may have some disadvantages for the mental health of humans. The robotic butler would allow an almost complete disconnect from the immediate environment and daily tasks. Yet, it is worth noting that the cognitive apparatus of human beings did not evolve for such a situation. The mind evolved in organisms that moved around, looked for food, partners, protection and shelter. In the case of humans that literally meant thinking on your feet as this was a necessity and dominant lifestyle more than 5000 years ago. If artificial intelligence releases people from all daily tasks, then we truly enter a new stage of human development.

References

Ahuja, A.S.: The impact of artificial intelligence in medicine on the future role of the physician. PeerJ **7**, e7702 (2019). https://doi.org/10.7717/peerj.7702

Amish Studies: https://groups.etown.edu/amishstudies/statistics/amish-population-profile-2020. 20 Nov 2020

Bennett, H.: Biometric mirror highlights flaws in artificial intelligence. https://about.unimelb.edu.au/newsroom/news/2018/july/biometric-mirror-highlights-flaws-in-artificial-intelligence. 24/07/2018. Retrieved 26 Nov 2020

Black, D.: A brief definition of mindfulness. Mindfulness Research Guide (2011). mindfulexperience.org. 9 Dec 2019

Boyle, M.: The Way Home: Tales from a Life Without Technology. Oneworld, London (2019)

Brown, K.W., Cordon, S.: Toward a phenomenology of mindfulness: subjective experience and emotional correlates. In: Didonna, F. (ed.) Clinical Handbook of Mindfulness, 59–84. Springer (2009)

Brown, K.W., Ryan, R.M.: The benefits of being present: Mindfulness and its role in psychological well being. J. Pers. Soc. Psychol. **84**(4), 822–848 (2003)

Chandler, S.: The Key To Successful AI: Hiding Its Use From People. Forbes. 29 Sept 2019

Chen, B.X., Metz, C.: Google's Duplex Uses A.I. to Mimic Humans (Sometimes). New York Times. 22 May (2019). https://www.nytimes.com/2019/05/22/technology/personaltech/ai-google-duplex.html. 18 Jan 2020

Curtis, C.: Scientists developed #MeTooBots to fight workplace sexual harassment—but we shouldn't rely on it. https://thenextweb.com/code-word/2020/01/08/scientists-developed-metoobots-to-fight-workplace-sexual-harassment-but-we-shouldnt-rely-on-it/. 09 Jan 2020

Dreyfus, H.L.: Why Heideggerian AI failed and how fixing it would require making it more Heideggerian. Philoso. Psychol. **20**(2), 247–268 (2007)

Feenberg, A., Heidegger: Marcuse and the Critique of Technology (2010). https://www.sfu.ca/~andrewf/Heideggertalksfu.htm. 28 Nov 2019

Fuchs, C.: Technik als Gestell bei Martin Heidegger. Eine Weise der Entbergung. http://cartoon.iguw.tuwien.ac.at/christian/technsoz/heidegger.html. 04 Dec 2019

Gorey, C.: AI 'mirror' that ranks attractiveness reveals tech's major flaw. Silicon republic 24 July 2018

Griffith, C.: Autonomous 'killer drones' being deployed in war zones. The Australian, 16 June 2020

Heidegger, M.: "Nur noch ein Gott kann uns retten," Der Spiegel 30 (Mai, 1976): 193–219. Trans. by W. Richardson as "Only a God Can Save Us". In: Sheehan, T. (ed.) Heidegger: The Man and the Thinker, pp. 45–67 (1981)

Heidegger, M.: Die Frage nach der Technik. Vortraege und Aufsaetze, 1949. Frankfurt a.M., Vittorio Klostermann, (2000)

Jigsaw: The challenge of identifying subtle forms of toxicity online. https://medium.com/the-false-positive/the-challenge-of-identifying-subtle-forms-of-toxicity-online-465505b6c4c9. 13 Dec 2018

Kryszczuk, M.D., Wenzel, M., Neo-Luddism: Contemporary work and beyond. Central and Eastern European Online Library. https://www.ceeol.com/search/article-detail?id=603964 2017

Longoni, C., Bonezzi, A., Morewedge, C.K.: Resistance to Medical Artificial Intelligence. J. Consum. Res. **46**, 629–650 (2019)

Luo, et al.: Frontiers: the impact of AI chatbot disclosure on customer purchases. Mark. Sci. **38**(6), 937–947 (2019)

Panch, T., Mattie, H., Celi, L.A.: The "inconvenient truth" about AI in healthcare. npj Digit. Med. 2, 77 (2019). https://doi.org/10.1038/s41746-019-0155-4

Rothman, J.: Is Heidegger contaminated by Nazism? The New Yorker, 28 Apr 2014

Sharr, A.: Heidegger's Hut. MIT Press, Cambridge (2006)

Snell, T., Bhullat, N.: Are cities bad for you? In: Psych: The Bulletin of the Australian Psychological Society Limited. 9–13 Dec 2019

The Luddites at 200. https://www.luddites200.org.uk/theLuddites.html. 20 Nov 2020

Thompson, E., Zahavi, D., Philosophical Issues: Phenomenology. In: Moscovitch, M., Zelazo, P. & Thompson, E. (Eds.), Cambridge handbook of consciousness. Cambridge: Cambridge University Press, pp. 67–87 (2007)

Wheeler, M., "Martin Heidegger", The Stanford Encyclopedia of Philosophy (Winter 2018 Edition), Edward N. Zalta (ed.). https://plato.stanford.edu/archives/win2018/entries/heidegger. 4 Dec 2019

Woolley, S.: We're fighting fake news AI bots by using more AI. That's a mistake. MIT Technology Review 08 Jan 2020

Chapter 6
Safety and Military Artificial Intelligence

A swarm of flying or swimming drones with lethal weapons and autonomous decision making is clearly within the reach of current technology. In principle, such a swarm could include thousands of robots and the technology represents a cheap solution to problems of warfare. Since no human combatants are involved, some may argue that it represents a safe option in military conflicts. However, there can be little doubt that weaponised AI has a psychological impact, with emotions such as fear, anxiety and guilt as consequences. This chapter introduces new forms of human control and discusses the modification of the current laws of armed conflict including the option to reduce any form of artificial superintelligence to human cognitive abilities. The concept of intrusion is introduced to discuss new forms of human-in-the-loop as a means to control AI systems.

6.1 Introduction

The use of autonomous lethal weapons systems that kill and destroy without human control is not new. These systems have been deployed by a number of countries for decades. "Loitering munition" exists in various forms and is sometimes referred to as suicide or kamikaze drones. The weapon system loiters around an area for some time, searches for targets and attacks once the presence is confirmed. In a crude form, the drone would detect a signal, for instance a radar signal, hone in on the target and then destroy it with the munition on board. The drone crashes on the target and is destroyed in the process. In this crude form, the process is fast and without any human control. Once the signal is detected, the location is destroyed. In more developed systems, there is selective targeting and the mission can be aborted. In a sense, loitering munitions are minefields that search and destroy.

The advantage of these systems is their fast reaction time; the disadvantage is that the human is completely out of the loop. A quick response is an obvious advantage

J. Diederich, *The Psychology of Artificial Superintelligence*,
Cognitive Systems Monographs 42,
https://doi.org/10.1007/978-3-030-71842-8_6

in the military context and AI systems can respond extremely quickly because they can process a vast amount of information efficently. As outlined in the introduction chapter, AI decision-making is magnitudes faster than human comprehension. This raises the prospect of "Flash Wars"—conflicts between two or more armies of weaponised AI systems that are over before any human can notice it (Franke 2018).

The argument in Chapter 1 that super-fast decision-making is not possible if natural processes have to be observed that develop at their own time scale, has limited application here. Intelligent loitering munitions are moved into place so that there is a fast reaction without any human input.

The nuclear conflict scenarios that terrorised worldwide populations in particular during the 1950s and 60s, are different to the situation explored here. Traditional nuclear conflict may include warning times and missile defence. In the scenario here, the conflict is over before any human knows. If the traditional nuclear conflict scenario is traumatising, then this new threat is even more so.

The instinctive reaction to the scenario above is to ask for international control and regulation of these weapon systems. However, it is extremely unlikely that artificial intelligence will be regulated or controlled worldwide. Even if there are ethical and regulatory standards in addition to policies for AI applications, it is very unlikely that these rules would be equally applied worldwide. It is more likely that there will be a continuing race, including collaboration and competition between nations and organisations, to develop ever more advanced forms of artificial intelligence. If the United Nations would endorse standards and every country would subscribe to them, there would still be a role for rogue players to use artificial intelligence to seek a niche dominance.

6.2 Attitudes Towards Military AI in the USA

The public's opinion on military artificial intelligence is important on a number of levels. Without public support there would be no funding and no development of the technology. Public opinion also influences decision makers when it comes to the deployment of the technology. Finally, in any military conflict there must be public support to sustain the effort and to offer resilience. Obviously, public opinion can change rapidly but it is still important to have a good assessment of the current views in order to estimate the psychological impact of the technology.

Morgan et al. (2020) report on two surveys in the USA to give a better assessment of the public's view on military AI and, in particular, autonomous weapons. Not surprisingly, the responses crucially depend on context. If anything is labelled as a desperate last resort of their own side, the public is more supportive of the measure.

Morgan et al. (2020, p. 102) summarise the results of the US survey as follows

> Defensive systems are more acceptable than offensive systems; using machines to attack machines is more acceptable than using them to attack people; and keeping a human in the loop in the use of lethal force is important. Finally, the respondents were surprisingly supportive of using AI to improve ISR (intelligence, surveillance & reconnaissance)

processing and decision making, but sharply rejected using it to falsely attack the reputations of foreign leaders.

Horowitz (2016) offers results on a survey on U.S. public attitudes regarding the use of military AI using Amazon's MTurk crowdsourcing platform (Figs. 6.1 and 6.2).

These two surveys demonstrate that the public may be in favour of controlling some forms of military AI. However, before the (limited) options for international control are discussed, the question must be asked if machines themselves can be responsible agents in an armed conflict.

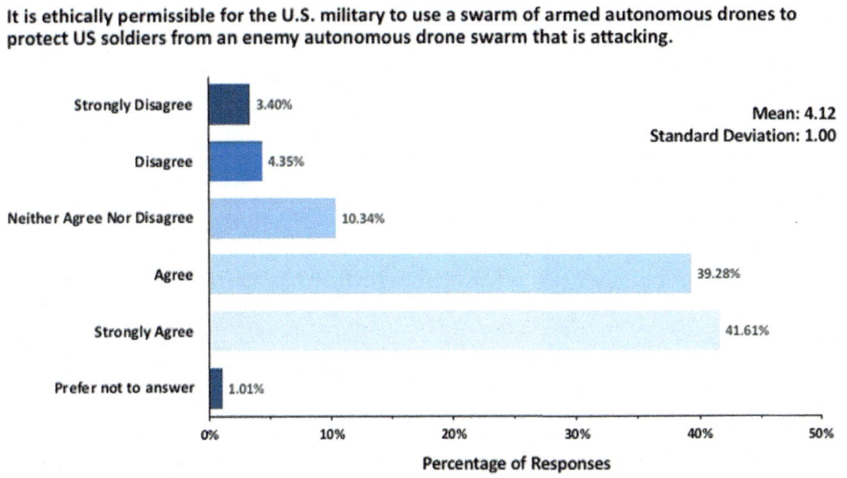

Fig. 6.1 Swarm of autonomous drones to protect US soldiers (in Morgan et al., 2020, p. 109)

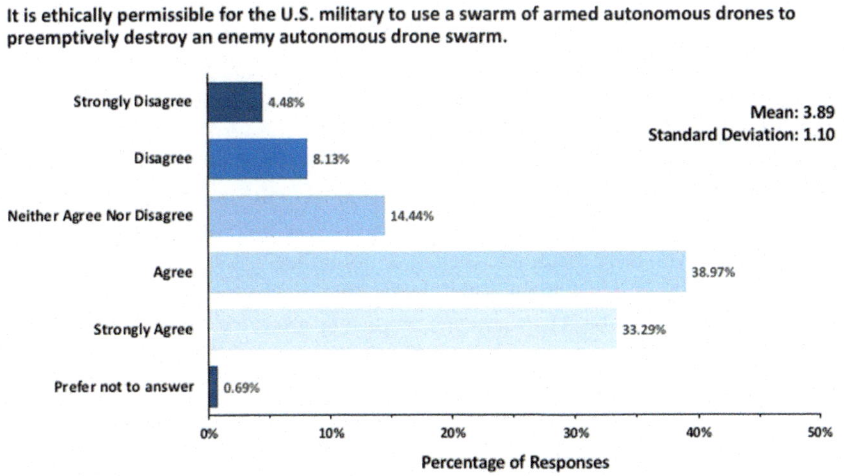

Fig. 6.2 Swarm of drones to destroy enemy drones (in Morgan et al., 2020, p. 109)

6.3 Are Machines Responsible Agents?

If a learning artificial intelligence system kills someone or injures a person, who is actually responsible? Who is liable in a criminal offence committed by a machine and who is punishable? Even an advanced AI system cannot be punished like a human being. For instance, there cannot be any jail time or other forms of sanctions as the AI system would not suffer due to a lack of personal freedoms.

These questions are currently being discussed and it is a less-than trivial topic. People have been injured by robots and self-driving cars already. John Kingston at the University of Brighton in the UK has investigated this legal field (Kingston 2020) in relation to criminal liability.

Criminal liability usually requires an action and a criminal intent. Kingston (2020) explores three scenarios that could apply to AI systems. The first case is called a "perpetrator via another" and applies if an offence has been committed by a child or an animal who is therefore considered to be innocent. Nevertheless, the person who instructed the innocent who committed the action can be held criminally reliable. If a dog owner instructs an animal to attack another person, the dog owner is responsible. If this scenario is applied to AI system, the machine could be considered an innocent agent and either the software designer, the programmer, the data analyst who conducted the machine learning or the user could be considered a "perpetrator via another."

The second scenario is called a "natural probable consequence" and this occurs when *the ordinary action* of an AI system is used inappropriately to perform a criminal act. The ordinary action of a self-driving car using artificial intelligence might, for instance, kill a person who is crossing the street. Assessing this situation is even more complicated: an avoidance action could also kill a person. As a matter of fact, an avoidance action may include a choice between people and those who suffer the most negative consequences. The question here is whether the designer or programmer of the self-driving car knew that killing a person was a probable consequence of using the car.

The third scenario is called direct liability, and this requires both an action and an intent. The action is easy to assess from a legal point of view as it would be (1) some behaviour that results in a criminal act or (2) the lack of an action to avoid the criminal activity (MIT Review 2018). However, the intend is the problem. If an AI system kills someone, who is actually responsible: (1) the owner of the system, (2) the operator of the system, (3) the designer or programmer, (4) the manufacturer or (5) the person who used machine learning to build the system?

Outside of criminal law, there is civil legislation in which case one of the relevant questions would be if the AI system is a product or a service. If it is a product, then according to John Kingston, product design legislation would apply (MIT Review 2018). If it is a service, then negligence comes into play. Here the question is if the responsible person, who cannot be the AI system, had a duty of care. The next question is if this duty was breached. Possible answers are complicated by the question of whether the designer, owner, or operator of the AI system could anticipate *all possible actions of the machine.*

In case of military applications, the case is more straightforward at first sight. AI systems that deploy anti-personal mines or operate machine guns are designed to kill and injure. In case such a military AI system is trained by machine learning, who is actually responsible for the killing: the commander who gave an order, the manufacturer of the machine, the software designer or the people who invented the machine learning algorithm?

6.3.1 Are Intelligent Killing Machines Responsible Agents?

The soldier who performs well in combat will be rewarded, either by a medal, some extended leave or a promotion. A soldier who performs well in combat, helps others and disregards her or his life is called a hero. Again, there are multiple rewards for this person, including fame. What is the name for a machine that kills well?

There is no such name and the question invites sarcasm or unfortunate attempts at humour. Obviously, we cannot anticipate a scenario where a machine is rewarded but we can imagine additional improvement cycles by the manufactures to make the machine even more lethal. If the killing machine is so good, who is actually responsible and contributed to this performance? And again, who is responsible for each and every individual death?

It is difficult to see that the designer or operator of a killing AI machine can anticipate all possible actions. If the machine learns fast and is deployed in new environments then it is difficult to see how actions could be anticipated. In the case of learning machines, it is necessary to ask again who has responsibility: the researchers who invented machine learning algorithms or knowledge representation systems, the designer of the robot, the people who deploy the machine or the person who gives the command to kill someone. Recently, employees of Google have voted not to participate in a military AI Project. Google is a software and data processing company with significant research and development efforts. Not many of the researchers would actually build killing machines. Their colleagues who voted against the military AI project, without being involved in the research directly, are even further removed from the act of killing. Nevertheless, these employees felt responsible enough to vote against any involvement in this project.

For many, it is an uncomfortable reality that conflicts are not between countries anymore but run deep through societies and communities. The enemy is a neighbour who may vote for a different party or expresses different thoughts on social media. This brings any conflict home and it even brings military conflict home. The "Slaughterbot" video by Stuart Russell and his colleagues demonstrates this point in a very powerful way.

The most uncomfortable outcome of this discussion is that *nobody is responsible* for the actions of killing machines. And this may even increase the universal solicitation of the technology. In the year 2020, people could mobilise against the dangers of advanced and autonomous artificial intelligence. While most AI researchers recognise the risk and the problems, only relatively few are outspoken about the dangers. Are all current AI researchers responsible for the killing machines of the future? Are tax payers who contribute to the relevant research responsible for the killing of the future?

6.3.2 Who Contributes to Killing Humans by Use of AI?

There is no single person who stands out as responsible if a *learning* AI system kills a human being. Drones can be deployed for a very long time and can learn at all times. Decision-making in combat can be so fast that no human being can exercise control. The commander who has given the order for search and kill may not be in complete control when the machine is actually in a position to act on the order. If there is no single human being responsible for the act of killing, then there must be a set of contributors and contributions.

This includes the designers of the machine, and the people who maintain it or develop systems for the upkeep of lethal machines. Further down the road, the contributions become minor or minimal and somewhere in the distance there is a non-zero number for Benjamin Franklin, the "inventor" of electricity. A scenario emerges that is quite unique and threatening: Nobody is really responsible and punishable for the killing of a human being.

The current grey areas when it comes to the laws of armed conflict are different to the future scenario that needs to be avoided at all cost. Under the current laws, a combatant who has surrendered and is a prisoner of war needs to be protected and not killed. The same prisoner of war may have committed war crimes the day before. Nevertheless, the law is clear-cut: this person should not be harmed. In late 2020, there is a discussion in Australia about the role of special forces during the long war in Afghanistan. It has been alleged that soldiers who served with honour and have received some of the highest awards, have also harmed or killed prisoners of war. The whole discussion surrounding the evaluation of guilt still focuses on the individual soldier who harmed a prisoner of war. It is this person who is primarily responsible, in addition to the officers who had oversight and governments that sent soldiers on multiple tours of duty. These people are not remote from the action as is the case with the manufacturer of a uniform. The difference is, of course, that a human being can be responsible for her or his actions unlike an animal, a child or a machine.

6.3.3 Nazi Atrocities, Liability and Personal Guilt

Oskar Groening was the famous accountant of Auschwitz, a young person with bookkeeping skills who joined a military fascist organisation during World War II. He was soon transferred to the concentration camp Auschwitz in Poland where he worked as a military clerk and also performed guard duties. Oskar Groening claimed not to have murdered a prisoner personally, however, he witnessed the mass killings and unbelievable cruelty towards children.

After World War II, Groening felt innocent and increasingly made public statements to confront Holocaust deniers. He reported that he felt some guilt but he was so sure of his innocence in a legal sense that he contributed to TV documentaries and gave a number of interviews. Very slowly, however, the interpretation of law changed in Germany, and eventually, not just the murderers themselves were charged but also people who contributed to the mass killings in support roles. Oscar Groening was charged and convicted, however, he did not serve a prison sentence since he died in his 90s before for the commencement of the jail term.

This is an example of how the legal and public views of guilt and liability changes over time. A person was considered innocent in a legal sense because there was no personal and direct connection to murder, but was convicted at a later point in time because of contributions to genocide. After World War II, there was an intense debate in Germany about the guilt and the role of almost everybody who lived at the time, with the sole exception of the persecuted and resistance fighters. This is an ongoing discussion in Germany many decades after World War II.

Oscar Groening serves as an example for an indirect contribution to mass murder, however, the distinction between those who conducted the killing and those who contributed still holds. What about other scenarios with no direct killing but with harmful or lethal outcomes for many and the contribution of most, if not all, of the global population?

6.3.4 Climate Change and Personal Guilt

Climate change kills too, directly and indirectly. Heat-waves can be deadly for the elderly and vulnerable in countries like Australia and the global South. Extreme weather events such as cyclones and hurricanes do not only cause loss to property but directly harm humans. An increase in infectious diseases is predicted and the worsening of air pollution will have a negative impact on humans (Wu et al. 2016).

Currently, most people contribute to climate change and very few can claim that they have a net zero effect. Clearly, there are differences between countries and continents, based on industrial production and also the release of carbon into the atmosphere by other means. Individual contributions are difficult to measure: a person may drive a petrol car and use commercial airplanes while at the same time invest into reforestation.

There is hardly any analogy here to the mass killing of people by use of weaponry, both in wars and mass genocides. These type of conflicts require tools and machines, from the primitive weapons of the stone age to the autonomous intelligent killing machines of today. In the past, machines of war were operated by humans; what is emerging now is qualitatively different: killing machines that learn and self improve without effective human control. This is different to anything that happened before and it is here already.

6.3.5 Non-humanoid Drones and Anxiety

There is no reason to believe that the killing machines of the future resemble human form. The android killers in movies such as "Terminator" or "Raised by Wolves" are close to the human form and create an 'Uncanny Valley Effect'. In other words, the closer the machines get to human form without reaching perfection, the creepier they are and the more entertaining the movie. But it is doubtful that the human form is optimal for military tasks.

The swarm of small drones in the "Slaughterbot" video is perfect for a mission that targets humans; inexpensive flying drones in a search-and-destroy mission with targeting based on well-defined features. These drones are certainly not in the uncanny valley because they resemble insects more than they resemble humans, but they trigger a different kind of fear: insectophobia or entomophobia.

The fear of insects is widespread and for good reasons. Insects carry diseases, they bite and can be poisonous. The fear of insects is possibly genetically programmed. Scientists at the Max Planck Institute for Human Cognitive and Brain Sciences (MPI CBS) in Leipzig and from Uppsala University in Sweden, recently made a crucial observation: Even in infants a stress reaction is evoked when they see a spider or a snake. This occurs already at the age of six months, when children have had little opportunity to learn that arachnids and reptiles can be dangerous (Hoehl et al. 2017). When the researchers showed pictures of a snake or a spider to the babies instead of a flower or a fish of the same size and colour, they reacted with significantly bigger pupils. Given the same light conditions, this change in size of the pupils points to the activation of the noradrenergic system in the brain, which is responsible for stress reactions. Even the youngest babies are stressed by insects and snakes. Hoehl et al. (2017) conclude that fear of snakes and spiders is of evolutionary origin and allows for very fast reaction times.

It may be possible to divert the attention of other dangerous animals but no such luck with insects. As stated earlier, it is not possible to argue with insects for the purpose of obtaining personal safety. And the larger the number, the more dangerous they are. As a result, humans are likely to react with fear to insect-like robotic drones, particularly when they come in swarms, due to evolutionary programming.

6.4 Emotion and Military AI

The Universal Declaration of Human Rights by the United Nation states that the *recognition of the inherent dignity and of the equal and inalienable rights of all members of the human family is the Foundation of Freedom, Justice and Peace.* The argument has been made that it is against the inherent dignity of humans to have life terminated by an autonomous AI system that makes an independent decision about life and death (Morgan et al. 2020, p. 34, Paass and Hecker, 2020, p. 417). Machines do not have moral qualities and no way of justifying actions with reference to moral belief systems shared by humans. Machines have no sense of existence and are not threatened by end of life issues. While it is possible to give belief systems to machines, this is fundamentally different to the sense of "being in the world" that humans have. In other words, a machine that cannot experience an end to life should not take away the life of a human being.

If somebody is sentenced to death within a country that has the death penalty, then this sentence is the result of a lengthy process and a judgement by human peers. In other words, a judgement by a group of people that can (and will) experience the end of life and in this sense are threatened by the termination of their being. However, a machine has a very different form of existence without any resource to moral judgement in the human sense and can make a decision in a split second.

Morgan et al. (2020, p. 34) point out that human emotion plays a very important part in war. It is possible to kill somebody in anger but there are also emotions such as guilt and regret that can moderate aggressive behaviour. Throughout history the whole spectrum of human emotion has been very important in warfare and it can result in the moderation of actions. Moral decision making and emotions, as experienced by biological systems, are unknown to machines. Hence, warfare by AI becomes more uncompromising and deadlier, and is indeed, a violation of human dignity.

6.5 Explanation and Military AI

If an AI system based on machine learning kills a human being, it would not "know" why. There is no way to interrogate the machine after the event and even with the most significant advances in the field of explanation and AI, it would not be possible to interview the device.

The rule extraction methods introduced in the chapter on explanation form rules based on combinations of input-output variables. The relationship between input features may be non-linear, and hence, the rules may be incomprehensible to humans. Even if hidden nodes in a neural network were to be decompiled by a decompositional rule extraction technique, there is no guarantee that this would result in information comprehensible to humans. There is simply no advancement at the horizon that would allow an interview with a deep learning network to explore its decisions and past actions. At least for the foreseeable future, deep learning systems are unaccountable by nature.

While swarms of armed AI systems exist and reportedly are being used in conflict, there is simply no way to debrief the machines the way this would happen with human pilots after a mission. It is not surprising that explanantion and AI (XAI) is high on the priority list of the military, particularly in relation to whether it performs as expected. This is described succinctly by the US General Paul J. Selva where he reportedly said (Morgan et al. 2020, p. 36):

> The entire enterprise of software engineers who are trying to build this artificial intelligence, deep learning space have not built the piece of software that can actually tell you what it's learned. And I think that's one of the milestones we're going to have to cross before we can actually get into a high-confidence area where we can say that technology is actually going to do what we want it to do, because not only can we physically test it; we can intellectually test it.

To complicate the matter further, in any conflict, the opponent would use data poisoning to feed learning machines wrong information. As a result, AI systems would be required to recognise trustful, versus untrustful, information with very high accuracy. It is not inconceivable that any opponent would specifically target the explanation component of AI applications (see Chapter 3). One way to do this would be to force the machines to accept very high dimensional feature spaces. The machine learning systems may then work with high accuracy, but the explanations would be poor and would result in the deterioration of trust in the performance of the machines. This in turn may result in periods when machines are not deployed due to quality assurance issues.

6.5.1 Explanation and Misinformation in Military Applications

When explanation components were developed for symbolic AI systems during the 1980s, a very important discovery was made. Initially, researchers in expert systems believed that the rules that "fired" (were used) during a problem-solving process could serve as an explanation (Gilbert 1989; Diederich 1992). After all, these rules explain how to get from the initial start state to a goal state in problem solving. For instance, the rules tell human observers how to start a game, how to play the middle section and how to win it. It was initially thought that this is all that is required to explain how to win a game.

It was soon discovered that some of the rules that fired during problem solving in an expert system are somewhat artificial to the human observer and were introduced to the knowledge base for purely technical reasons. These artificial and counter-intuitive rules tend to confuse people and destroy the entire explanation attempt. So, something else was required; something that is not necessarily a complete and close representation of the problem-solving process of the AI system but a closer match to human understanding. As a result, symbolic rule sets and knowledge bases where built for explanation purposes only.

The next step was the realisation that the rules required for explanation must not have any relationship with the internal functioning of the AI system at all. For instance, people accept explanations for physical processes that are far removed from reality. Not too long ago, by way of example, an accepted explanation for thunder was that the gods were angry. This perfectly understandable and intuitive explanation was accepted for centuries. And we are clearly not out of the woods yet when it comes to the utility of these types of explanations. For a century, researchers have struggled to explain quantum mechanics and used imagery and analogies that specifically appeal to the human understanding (e.g. "Schroedinger's cat"). The wider public appreciates these efforts in order to understand what is clearly a complicated topic involving very counter-intuitive phenomena (e.g. entanglement).

It is therefore conceivable that a military opponent may hack defence systems in order to provide intuitive explanations that are factually wrong. The hacker may use results from cognitive psychology, and in particular, biases in reasoning to make explanations as intuitive as possible.

Tversky and Kahneman (1981) are famous for their research into human decision making and cognitive biases. Tversky and Kahnemann explored if humans are truly rational in their decision-making. They challenged the assumption that rationality was a guiding principle in both economics and psychology. The following experiment from their 1981 paper explains cognitive biases in human decision making, specifically the acceptance of risk and the avoidance of loss. An experiment (Tversky and Kahneman 1981, p. 453) that appears timely from the perspective of 2020 is described here:

> Problem 1 [N = 152]: Imagine that the U.S. is preparing for the outbreak of an unusual Asian disease, which is expected to kill 600 people. Two alternative programs to combat the disease have been proposed. Assume that the exact scientific estimate of the consequences of the programs are as follows: If Program A is adopted, 200 people will be saved. [72 percent] If Program B is adopted, there is 1/3 probability that 600 people will be saved, and 2/3 probability that no people will be saved. [28 percent] Which of the two programs would you favor? … The majority choice in this problem is risk averse: the prospect of certainly saving 200 lives is more attractive than a risky prospect of equal expected value, that is, a one-in-three chance of saving 600 lives.

These principles can be used to explain to human users that the decisions of an AI system are compassionate and considerate while this is not the case when another frame of reference is used. Clearly, certain explanations are better than others and those consistent with human cognitive biases may be preferred.

Conspiracy theories are alternative explanations for events that are both inconsistent with recorded facts and not shared by a majority of the population. Typically, conspiracy theories attribute certain events to hidden organisations or supernatural forces. Classic conspiracy theories focus on the moon landing or the death of John F. Kennedy, however, since the introduction of social media, there is a true proliferation of conspiracy theories.

Currently, many conspiracy theories are used in the social media sphere to distract and confuse. It is entirely conceivable that AI systems can be used to generate conspiracy theories that appeal to human cognitive biases. Let's use one current conspiracy theory as an example: there is no coronavirus and lockdowns have been introduced worldwide by "deep state" actors to enforce behaviours that mitigate the effects of climate change. This theory is counterfactual since the existence of the virus has been proven, and furthermore, it appeals to those who like to attribute events and outcomes to hidden organisations. The theory also invokes cognitive biases since both the coronavirus and climate change are dangerous to human life and humans are biased to prevent the loss of what is most precious. The plausibility of the theory is enhanced since both explanations (lockdowns helps to prevent the transmission of the coronavirus and lockdowns mitigate climate change) appeal to the bias to reduce the loss of human life.

While nuanced conspiracy theories are currently being introduced on social media platforms to distract and confuse, these theories can also serve as powerful explanations in military conflicts. If AI systems are hacked, then conspiracy theories can be inserted in explanations in order to confuse or distract the opponent.

6.6 The Rules of Armed Conflict

Even though international regulation does not sound promising, a look at the current "Rules for Armed Conflict" which are part of international law is helpful. The International Red Cross states as part of a *training manual for soldiers* (The Law of Armed Conflict, ICRC, https://www.icrc.org/en/doc/assets/files/other/law1_final.pdf):

> The law of armed conflict is a branch of international law, the law that States have agreed to accept as binding upon them in their dealings with other States. As well as governing relationships between States, international law applies to the conduct of hostilities within a State. It is therefore crucial to the profession of arms – your profession … The law is therefore quite simply your State's law. It is binding on you as a member of the armed forces of your State.

There are a number a number of core concepts in the Law of Armed Conflict. The *training manual for soldiers* (The Law of Armed Conflict, ICRC) outlines the following six concepts, which will be unpacked further in the next section in terms of armed conflict and AI:

> **Distinction**: Every soldier must clearly distinguish between combatants and civilians or the civilian population as such. Combatants may be attacked unless they are out of action. Civilians are protected from attack but lose that protection whenever they take a direct part in hostilities for the time of their participation. Similarly, soldiers must always distinguish between military objectives which can be attacked and civilian objects which must be respected.
>
> **Proportionality**: When military objectives are attacked, civilians and civilian objects must be spared from incidental or collateral damage. Excessive use of force quite clearly violates the law of armed conflict … Good planning and clear rules of engagement are required to stay within the law.
>
> **Military Necessity**: This principle is enshrined in the preamble to the 1868 St. Petersburg Declaration, which states that "the only legitimate object which States should endeavour to accomplish during war is to weaken the military forces of the enemy" and that "for this purpose it is sufficient to disable the greatest possible number of men". Today it should also include women.
>
> **Limitation**: In any armed conflict, the right of the parties involved to choose methods and means of warfare is not unlimited. Weapons and tactics that are of a nature to cause unnecessary suffering or superfluous injury are prohibited (e.g. anti-personnel mines).

Good Faith: This is a customary principle of warfare. The military should show good faith in their interpretation of the law of armed conflict. Good faith must also be observed in negotiations between opponents and with humanitarian organizations.

Humane Treatment and Non-Discrimination: All people must be treated humanely and without discrimination based on sex, nationality, race, religion or political beliefs. Those who are out of action (hors de combat), such as surrendering combatants, air crew parachuting from downed aircraft, the wounded, sick and shipwrecked, prisoners of war and other captives and detainees, must be identified as such and treated humanely.

6.6.1 The Rules of Armed Conflict and AI

Anybody with a TV set and a superficial interest in history knows that the laws of armed conflict are not observed in reality. Civilian populations have been targeted on a regular basis, the carpet bombing of densely populated areas did occur in the twentieth century, in addition to the destruction of forested and agricultural areas. In light of these well-documented events, the laws of armed conflict almost seem cynical. So why then consider this part of international law in the context of very advanced military artificial intelligence?

The laws of armed conflict are a set of rules and machines follow rules. Here is a minuscule chance that an international agreement about the role of intelligent autonomous systems in armed conflict can be programmed directly into machines. If this would occur, then an opponent who is unarmed would simply be left alone. Intelligent, armed machines would not intrude into areas where people cannot represent a threat due to the absence of arms and limited utilisation of technology (for instance, areas settled by the Amish people in the United States and Canada or remote aboriginal communities in Australia).

Currently, soldiers and commanders must follow the rules of armed conflict as described in the six core concepts above, even when frightened and under direct threat. Naturally, combat and the denial of resources can result in extreme mental states including confusion, high arousal, anger and anxiety. As a result, rules are often disregarded in the "fog of war". By contrast, AI systems are unaffected by emotional states, and hence, there is a (very) theoretical chance to program this set of rules directly into machines. This is easy by use of direct programming but can be achieved with machine learning as well. The case can be made here for hybrid symbolic/machine learning systems that combine the advantages of direct programming and the adaptability of machine learning.

6.6.2 An Extended Set of Rules for Armed AI Conflicts

The current rules of armed conflict make sure that the "Slaughterbots" in Stuart Russell's video are illegal, both in conflicts between states and non-state actors. If there is an all-out war including artificial intelligence, then obviously all parts of the population would be affected. This very much includes individuals and groups who have decided to use technology to a lesser extent. It may be the case that these groups are targeted less because they represent no or a very limited threat, but there is no escaping the impact of military AI in case of war, even in the remotest corners of the word. Consequently the militarisation of artificial intelligence is a threat to each and every human on this planet.

Hence, it is absolutely necessary that there are rules of war similar to the Geneva conventions' that regulate and even prohibit forms of advanced artificial intelligence. As stated above, these rules and regulations may be discarded by rogue players but there must be generally accepted norms. The most important rule is that autonomous killing machines without human decision making must be banned at all cost. The decision to kill or not must not be made by a machine, there must be human control and every possible safeguard. Secondly, any decision about escalating conflict by allocating additional resources, must not be made by a machine alone. Algorithms based on artificial intelligence have created divisions and escalated conflicts in social media, and similar processes must be avoided in the case of war. Thirdly, non-combatants who do not pose a threat due to limited technological means must be allowed to stay out of conflict and should be protected.

Explanation in artificial intelligence is crucial due to safety considerations. If there is warfare with artificially superintelligent machines, then these devices must *explain their actions at all points in time*. This takes into consideration that any conflict between artificially intelligent devices may be incredibly fast by human standards. Explanation in artificial intelligence typically slows processes down, mostly due to the complexity of the underlying reasoning processes, such as reasoning about the state of belief of other combatants. In the struggle for human control, this slowdown of decision making in war can only be welcome.

The RAND Corp. published a taxonomy of AI risks with regard to military applications (Fig. 6.3).

Ethical	**Operational**	**Strategic**
Law of Armed Conflict Accountability and Moral Responsibility Human Dignity Human Rights and Privacy	Trust and Reliability Hacking, Data Poisoning, and Adversarial Attacks Accidents and Emergent Risks	Thresholds Escalation Management Proliferation Strategic Stability

Fig. 6.3 Taxonomy of artificial intelligence risk. Morgan et al. (2020, p. xiii)

6.7 Human Control of Advanced AI

6.7.1 Human AI Partnerships

One way to mitigate the risk of artificial intelligence within military situations is currently being investigated by the military. This includes pairing a human operator, a soldier, with an autonomous artificial intelligence machine. The pairing needs to be permanent so that both agents, the human and the AI system, can learn from each other. One scenario includes a human soldier and a robotic system that carries equipment, ammunitions and weapons, and potentially operates arms under human control.

The scenario is similar to the pairing of a soldier and dog, which has been used for centuries. The human soldier supervises the machine and exercises control. The soldier has the option and the means to shut down the machine completely if necessary. Since both are close to each other in the same operational theatre, the risk of uncontrolled harm by the AI system is significantly reduced. In the event the human operator is disabled, fatigued or loses oversight of the situation, the element of control is reduced or eliminated through the support of the AI.

In the case of human AI partnerships, where a robot is paired with human operator, an obvious requirement is that the machine operates at a time scale that allows control by the human partner. If the decision-making of the machine is significantly faster compared to human cognitive abilities, then there would be effectively no or reduced oversight. Hence, the processing speed of the machine must match human capabilities. The partnership slows the machine down and this increases the chance of human control.

6.7.2 Intrusion

In the context of this book, intrusion means that a human operator takes over the actions of an AI system. The human literally replaces the AI system at least for a period of time. During the period of intrusion, the human may continue to respond to inputs from the environment but may also inspect past behaviours of the machine and review software as part of quality assurance. If the human replaces the actions of an AI system, then obviously actions may not be as fast as the normal response time of the machine.

Using this definition, intrusion is a completely normal process in many AI applications and is not limited to machine intelligence at all. Financial analyst, for instance, may stop a trading AI system for a period of time to review its actions. Almost always, this results in the adjustment of parameters and software. Once the system has been tested, it is deployed again.

In the airline industry, aircraft are frequently taken apart for maintenance purposes. Individual components are tested and potentially replaced. The entire ensemble of components is tested again, and once it meets certain standards, the machine is operational again. This process of inspection and testing is a cornerstone of many engineering applications. Nevertheless, in the context of AI safety, intrusion goes further.

Figuratively speaking, intrusion means that a human operator is slipping into the AI system to take over its functions. Again, this requires that human behaviour and the actions of the artificial intelligence system operate roughly at the same time scale. If the decision making and the actions of the AI system is a million times faster than human perceptive and cognitive ability, then there is no chance for human control. Since machines can communicate at the speed of light and electronic devices are very fast compared to biological systems, there is a mismatch that eliminates human control unless processing speeds are adjusted. If this is not the case, and there are no restrictions on the AI system, then there is effectively no control by humans.

During intrusion the human operator takes over from the machine and reviews past actions. In addition, the human learns from the machine and vice versa. For the discussion here, this *excludes* a merger between brains and machines as is currently in development in the field of brain-machine interfaces. As outlined in the chapter on Transhumanism, the merger of minds and machines has consequences that cannot even be anticipated yet from a psychological point of view. Basically, there is no human control if the human is also a machine. The definition of intrusion here is *very conservative*, it literally means a human operator, described here as a person in the year 2020, taking over the operation of a machine.

Critics may argue that this limits the cognitive abilities of a machine to that of the cognitive ability of humans. This is very much the case and, indeed, the intention here. If this were not the case, runaway processes of self-improvement in artificial intelligence would be possible without any form of human control. There should be no singularity, as discussed in Chapter 1.

6.7.2.1 Intrusion and Digital Clones

The chapter on digital clones introduced the idea of an artificial intelligence system that closely mimics the professional activities of a working individual. The chapter also describes the Live Training option whereby the human original takes over from the chatbot and answers all questions of the users as required. In this case, the human literally replaces the bot and interaction is used to create software that can be used in future development cycles of the system. Naturally, a human taking over the actions of an AI system, or even replacing the system, represents a form of control.

Let's use the idea of digital clones as described in the previous chapters as an example for implementing safety measures and exercising control over an AI system. Whenever the human original interacts with users, the AI system learns. If this happens repeatedly, the AI system would re-adjust its behaviour and model the human original more closely. In this context, it certainly helps to use machine learning algorithms that realise one shot learning or can learn on the basis of relatively few training examples so that the human original must not generate millions of data points. In the best of all worlds, learning by instruction would be used so that every action by the human original automatically represents a rule that *must be observed* by the AI system.

In the case of digital clones, the clone must allow the intrusion of the human original to take over its actions. Clearly, this idea breaks down if there are a large number of digital clones that model the human original. At least in principle, there could be any number of clones and no way for the human original to take over from (and for) all of them, and by doing so making sure that they resemble the behaviour of the original.

Limiting the number of digital copies of the human original is artificial and clearly will not work in practice. In a military environment, it may be beneficial to have a very large number of copies in principle and there would be no inclination to limit the number in advance. As a result, additional security measurements would be necessary.

One possibility of such security would be changing the code and the knowledge sources that the clone is built upon, randomly, step-by-step, until the machine becomes useless. For instance, it should be possible to encrypt part of the code of a clone and by doing so severely limiting its functionality. Only if the clone allows the intrusion of the human original from time to time, the encryption is reversed. If there is no intrusion, the clone becomes useless over a brief period of time.

Processes like these make sure that the clone always matches the original and machine learning is used to facilitate the approximation of behaviour between clone and original, at least within small AI systems.

6.7.2.2 Intrusion for Large AI Systems

The introduction chapter provided definitions of artificial superintelligence that very much included systems build on current, mainstream technology. This includes very large systems and one of the examples is China's integrated AI-driven social credit system. Clearly, for systems of this nature, intrusion by one human is not possible. In this case, task decomposition is required. The entire system is broken down into individual tasks and functions that can be managed by a human operator. People involved in this process of intrusion can co-operate, however, the main constraint is that processing capabilities cannot go beyond human comprehension. This is a very severe limitation, and it is important to point out what this ongoing reference to human cognitive ability actually means.

The human working memory is a good example with which to explore this issue further. The human working memory can hold a limited number of items for a limited period of time. These items may differ in nature, and the only way to increase the capacity of working memory is by "chunking": items are combined so that there are a limited number of chunks to hold in working memory. This is an almost ridiculous limitation compared with the rapid access memory (RAM) of modern processors. Yet these limitations are a necessary requirement just to have some element of human control.

6.8 Artificial General Intelligence and Military AI

A central theme of this book is that artificial superintelligence, or any form of advanced artificial intelligence, has inescapable consequences for everybody. Many of these consequences are psychological in nature. Even people who do not use the technology are affected, since they are exposed to the solicitation of AI, i.e. the temptation to use the superintelligence for any purpose. Even though the risks associated with advanced AI systems have been highlighted by researchers and even celebrities, there is practically no public discussion compared to the big topics of the day such as climate change. In the absence of any significant public debate, we are all directly or indirectly contributing to the development of an AI superintelligence—either as researchers, investors, practitioners or simply as taxpayers.

A natural reaction would be to say that it is time to take a breath, maybe pause certain developments, and have a discussion of the possible consequences. But this is not possible since the militarisation of artificial intelligence is progressing with an arms race between countries and organisations. In 2017, President Putin of Russia said that the nation that leads in artificial intelligence will be the "ruler of the world." Putin said that "artificial intelligence is the future, not only for Russia, but for all humankind" according to the news agency RT. President Putin predicted that future wars will be fought by countries using drones. "When one parties' drones are destroyed by drones of another, it will have no choice but to surrender" said the Russian president, according to the Associated Press.

At the time this book is written, people are not out on the street protesting a future with drone wars. So it is necessary to contemplate all options for a safer future, including those that call for international regulation of drone-based military war.

One such option is to insist on artificial general intelligence in military applications. Artificial general intelligence (AGI) is a hypothetical form of machine cognition that has the capacity to understand and learn any task a human being can. Essentially, the idea here is to make the hardest problem in artificial intelligence not only a priority but a condition; the first problem that needs to be solved before machines are even considered that harm humans without any direct control by a person. This is not meant to be a justification for the use of AI in military applications, but instead a requirement of general AI would most likely delay the deployment of AI systems in defence applications by decades, if not longer. It would allow for time to reflect on and discuss the implications of the technology.

Artificial general intelligence is sometimes referred to as strong AI, full AI or general artificial intelligence. Sometimes, the requirements for artificial general intelligence include some form of consciousness and self-awareness. Current forms of artificial intelligence are weak AI, or narrow AI, since relatively narrow problems are solved only, even though this can be achieved with superhuman performance.

There is currently no agreed definition of "artificial general intelligence". Since the general objective is ill-defined, there is also no way to measure progress. However, there appears to be general agreement that a form of artificial general intelligence must be able to solve the hard problems such as language and speech comprehension. This includes metaphors, analogies and neologisms that people freely introduce into their language. Native speakers within a language community have very little problem understanding these new words. However, for an artificial intelligence system, the appropriate comprehension of metaphors and analogies is among the hardest problems (Feldman, 2008).

Researchers have argued that any system with full language comprehension ability must be embodied or have knowledge representations that refer to bodily movements and experience. For instance, metaphors frequently include physical activity and there needs to be an embodied, physical experience to fully understand these actions. The expression "His heart jumped for joy" is a metaphor since hearts do not jump in reality. Yet anybody who knows the physical experience of jumping and the emotional sensation of joy instantaneously understands this metaphor.

Artificial general intelligence systems must have the capacity to understand metaphors with ease. As central components of human language comprehension, they are also important for communication in the military. In the context of weaponised AI, the requirement for artificial general intelligence must go significantly beyond the capacity to understand human communication.

An artificial general intelligence system in a military context must have the capacity to understand physical threat and end-of-life issues. It must understand, if not experience, pain and all the sensations of being wounded. The AGI must understand death, the threat of being killed, the experience of a good friend passing away, and that the end of human life has an impact on a community. The AGI must further understand trauma and the long-term impact of combat on human lives.

This list of requirements for an artificial general intelligence in a military context is in no way complete. There are many more aspects of human life when it comes to being threatened, harmed and experiencing the end of life. Only when an AI system has a human level of comprehension, and possibly experience of these issues, can a machine be in a position to make any decisions about human life.

Nobody knows how to build such a machine, and there is very little agreement on the criteria for success. Yet, for the ethical reasons outlined above, such a machine would be a minimum requirement for deployment in military applications.

References

Diederich, J.: Explanation and artificial neural networks. Int. J. Man Mach. Stud. **37**, 335–355 (1992)

Feldman, J.: From Molecule to Metaphor: A Neural Theory of Language. The MIT Press, Boston (2008)

Franke, U.E.: Flash wars: Where could an autonomous weapons revolution lead us? Making it harder to attribute an attack to a perpetrator will have repercussions beyond non-state use. European Power, Commentary (22 Nov 2018)

Gilbert, N.: Explanation and dialogue. Knowl. Eng. Rev. **4**(3), 235–247 (1989)

Hoehl, S., Hellmer, K., Johansson, M., Gredebäck, G.: Itsy bitsy spider…: Infants react with increased arousal to spiders and snakes. Front. Psychol. **8**, 1710 (2017). https://www.cbs.mpg.de/Fear-of-spiders-and-snakes-is-deeply-embedded-in-us (19 Oct 2020)

Horowitz, M.C.: Public opinion and the politics of the killer robots debate. Res. Polit. **3**, 1 (2016)

https://www.icrc.org/en/doc/assets/files/other/law1_final.pdf

Kingston, J.: Artificial intelligence and legal liability. https://www.researchgate.net/profile/John_Kingston2/publication/309695295_Artificial_Intelligence_and_Legal_Liability/links/5a39397caca27208acc79e70/Artificial-Intelligence-and-Legal-Liability.pdf (13 Oct 2020)

Morgan, F.E., Boudreaux, B., Lohn, A.J., Ashby, M., Curriden, C., Klima, K., Grossman, D.: Military Applications of Artificial Intelligence Ethical Concerns in an Uncertain World. Published by the RAND Corporation, Santa Monica, CA (March 2020)

Paass, G., Hecker, D., Kuenstliche Intelligenz.: Was steckt hinter der Technologie der Zukunft? Wiesbaden, Springer Vieweg (2020)

Tversky, A., Kahneman, D.: The framing of decisions and the psychology of choice. Science **211**, 453–458 (1981)

When an AI finally kills someone, who will be responsible? MIT Technol. Rev. https://www.technologyreview.com/2018/03/12/144746/when-an-ai-finally-kills-someone-who-will-be-responsible/ (12 Mar 2018)

Wu, X., Lu, Y., Zhou, S., Chen, L., Xu, B.: Impact of climate change on human infectious diseases: Empirical evidence and human adaptation. Environ. Int. **86**, 14–23 (2016)

Chapter 7
Risks to Artificial Intelligence

Why do people insult, harm or destroy artificially intelligent machines? Why is the most vile abuse directed at chatbots and why are conversational agents trained by users to become racist? This chapter explores intentional and unintentional abuse of artificially intelligent machines as well as the physical damage inflicted on them. Interestingly, some of the areas have been explored already in those chapters that discussed the risk of artificial intelligence to humans. For instance, multiple cheap drones are used in flying weaponised swarms because the loss of a single one of these machines does not matter much. AI robots are sometimes designed to tolerate physical damage by humans, for instance, machines that teach social skills to autistic children. But why must intelligent beings that are considered to be sub- or non-human tolerate so much abuse?

7.1 Introduction

The previous chapters touched on some of the areas where an advanced artificial intelligence is a risk to humans. In contrast, this chapter focuses on *risks that humans pose to an artificial superintelligence* or any form of AI. This includes applications where the risk to the AI, and even damage or destruction of the machine, is intentional. Intelligent robots can work in environments that are dangerous to humans. In these applications, damage to the machines and reduced performance is anticipated, and due to safety reasons, any degradation in performance is limited by design to avoid complete failure and loss of functionality. Robot AI systems also work in service roles including the protection of property.

Recently, researchers from a number of Japanese universities monitored a shopping centre in Osaka by use of a remotely operated robot (Darling 2015). If the robot detected somebody in its path, it would politely ask the person to step aside. If the person would not respond, the robot would move away and would choose a different

J. Diederich, *The Psychology of Artificial Superintelligence*,
Cognitive Systems Monographs 42,
https://doi.org/10.1007/978-3-030-71842-8_7

direction. The robot quickly attracted the attention of groups of children who would play with the robot when unsupervised. The children gave the robot meaningless instructions and frequently tried to block its path.

Sometimes a child would step aside when asked by the robot and then quickly move into its path again. Sometimes the behaviour of the children would escalate and they would start hitting and kicking the robot. In addition, the children would call the robot bad names.

The researchers analysed this bullying behaviour and developed a computer simulation of the children's abuse. The behaviour would mostly occur when the children were in groups and when there are no adults in the vicinity (Darling 2015).

The researchers then developed algorithms so that the robot could avoid situations where humans potentially would pose a threat to the robot. The algorithms instructed the robot to run away from humans below a certain height and to run in the direction of taller people. In this case, people below 1.4 m were identified as dangerous by the robot and people above 1.4 m, who move in groups, were classified as safe. The idea was that a tall person in the company of others would potentially protect the robot from the abuse by tiny people (Darling 2015).

The researchers then tried to find an answer to the question of why children abuse robots. Interestingly, the majority of children described the robot as human-like and only a minority of 13% described it as machine-like (Darling 2015). Half of the children believed that their behaviour was painful to the robot. Obviously, this raises the question of why a robot, described as human-like by the majority of children, would attract abuse?

Naturally, it is not just children who hurt robots. It is a wider phenomenon and particularly directed against robots that resemble humans. What is the reason for this aggression? Could it be a fear of robots taking human jobs or maybe replacing humans in society?

On social media, videos of people attacking robots are very popular (Bromwich 2019). One explanation is that this is an us-versus-them phenomenon, an extension of an age-old tribal tendency that directs aggression against outsiders and invaders. There is a theory that the tendancy to humanise robots arises from the trend to anthropomorphise them (Bromwich 2019). For instance, people have a tendency to give names to robots, either in the workplace or at home, even if the robot does not come with a given name (e.g. a brand name). Furthermore, since science has taught humans that they are nothing but biological machines, the robot is just the slightly different thing that is being rejected when it comes too close. First, people attribute human features to robots and then they reject them when they come too close to humanity.

Below is a brief summary of why people abuse robots:

1. *An us vs them psychology.* Tribal instincts are directed against the machine that looks almost human. It is the fear of the unknown but also an action directed against something that is seen as an invader.
2. *Robots as competition.* Robots represent advanced automation which, in turn, means job losses. Robots can perform low-skilled jobs which can be seen as

a threat by people normally performing these jobs. There are no means to resist progress and automation, and hence, aggression is a way of expressing frustration with the ever-narrowing choices that a low-skilled population has.

3. *Robots are seen as emotionless and lacking empathy.* Industrial or military robots, in particular, perform tasks without any expression of emotion, frustration or empathy. Currently, it is not really possible to talk to a robot socially and to engage a machine, for instance, as a way to express concerns about the way a certain job is done. If there is a conflict between people, the mature way is to come together and to find a solution. This negotiation requires a good understanding of the emotional and mental state of the other, and in the case of the robot, it is simply assumed that there is no ability for it to empathise with others. Hence, human types of problem solving and resolving conflicts are simply not available when interacting with a machine.
4. *Robots are creepy.* In particular, robots that are almost perfectly human in appearance are seen as eerie. The Terminator films serve as a good example: A superintelligent machine is featured, who looks almost human and has superhuman capabilities, such as the ability to time travel. The Terminator is an emotionless killing machine. From a commercial and social-learning perspective, people pay a lot of money to see humans fighting an artificial superintelligence like the Terminator. Such violence is experienced as satisfying for the observer. Indeed, a whole section of the movie is devoted to such violence, and similarly, the computer games industry lives on a simulated conflict between humans and those who are simply creepy (robots, zombies, half-dead humans etc.).
5. The chapter on Neo-Luddism in this book outlined a general critique of technology and how it applies to artificial intelligence. Heidegger emphasised that technology is not just a tool but that it *changes the way the world is experienced.* This is the case even if people have no direct hands-on experience with the technology. In a similar vein, it is possible that people sense that the presence of a robot changes their work and life experience fundamentally. If the robot is a "colleague" who is working tirelessly for 24-hours, then the presence of the machine changes the work experience fundamentally and irreversible. Whatever work conditions existed before this is lost and there is no known way of going back. People may sense the significance of the change with the introduction of robots into the work environment, and may well react with rejection and even violence.

So intelligent and state-of-the-art robots that appear to be close to perfection when it comes to resembling humans are a particular target for violence and abuse. Machines may be super intelligent and perform very important tasks, yet there is an instinctive reaction against them. Why are machines that are superhuman in some areas, including logical reasoning and learning, considered subhuman in other areas, in particular with regard to emotional capabilities and feelings of empathy?

7.2 The Uncanny Valley of Humans and Machines

The uncanny valley hypothesis was introduced by Masahiro Mori 50 years ago. The concept is important in human computer interaction, artificial intelligence and robotics. Maori observed that the *attractiveness of robots does not increase linearly* as they become more human-like. At first, users find a more human appearance of a robot appealing but the attractiveness dips as the robot approximates the human form more closely. Mori (2012, p. 98) writes

> I have noticed that, in climbing toward the goal of making robots appear like a human, our affinity for them increases until we come to a valley …, which I call the uncanny valley.

Robots that are "almost" human appear to be creepy and the effect only vanishes as the machines approximates a perfect human appearance. The uncanny valley is this dip in attractiveness prior to "being fully human". It is important to note that the reference point in this context, the human, is a healthy and fully functional individual.

Mori (2012, p. 99) elaborates, for instance, on an unexpected response one might have to shaking hands with someone who has a prosthetic hand. He writes

> For example, we could be startled during a handshake by its limp boneless grip together with its texture and coldness. When this happens, we lose our sense of affinity, and the hand becomes uncanny. In mathematical terms, this can be represented by a negative value. Therefore, in this case, the appearance of the prosthetic hand is quite human like, but the level of affinity is negative, thus placing the hand near the bottom of the valley.

Interestingly, the uncanny valley concept also applies to user interfaces and not just robots. It includes chatbots or other human-like software realisations that aim to facilitate user interaction. Machines and programs in the uncanny valley can trigger avoidance and rejection. The uncanny valley hypothesis has been offered as an explanation for why users abuse machines. If a chatbot appears to be human like, but the natural language processing component is not up to scratch, then users can be inclined to be hostile towards the bot.

7.2.1 Challenging the Uncanny Valley Concept

It is generally believed that people find humanoid robots more appealing than machines without human-like features. It is sometimes assumed that the underlying psychological process includes "adding a mind" to the machine due to its appearance. This is sometimes referred to as the "mind-perception theory". As we have seen above, this only works up to a point, since machines that resemble humans too closely are rejected. Recent research points to a process whereby first a mind is *added to the machine*, which is followed by the realisation that clearly this thing is not a human, so that the *mind hypothesis is withdrawn or neutralized* (Wang et al. 2020). In other words, there is the realisation that there is a machine that appears to be human but it is a device only, and hence, is being rejected. Hence, there is not a single one-shot process whereby a mind is added to the machine in the perception of the observer, but rather, there is a dynamic process which includes first adding the mind and then withdrawing the hypothesis.

In some human conditions, such as extreme autism or some psychotic disorders, there is a mechanism called mind-blindness: the inability to distinguish between humans and machines.

7.3 The Uncanny Valley and Autism Spectrum Disorder

Does the uncanny valley hypothesis apply to humans as well? Individuals with Autism Spectrum Disorder (ASD) can appear robotic with a lack of facial expression, the use of repetitive language and a limited repertoire of gestures. There appears to be a lack of empathy even though there can be no doubt that individuals with autism spectrum disorder experience a full range of emotions. Nevertheless, some of the workplace problems that individuals with high functioning autism report are often attributed to a limited range of social interactions and the lack of interest in anything that is not factual and evidence-based. Is the uncanny valley phenomenon behind the rejection and bullying that individuals with ASD frequently experience?

As we are now broadening the uncanny valley concept and talk about a very human condition, it might be useful to briefly review the diagnostic criteria for ASD according to DSM-5 (The Diagnostic and Statistical Manual of Mental Disorders, 299.00, F84.0. Please note the diagnostic criteria are abbreviated here):

1. Persistent deficits in social communication and social interaction across multiple contexts, as manifested by
2. Deficits in social-emotional reciprocity, from an unusual social approach and failure of normal back-and-forth conversation, a reduced sharing of interests, emotions and a failure to initiate or respond to social interactions.
3. Deficits in non-verbal communicative behaviours used for social interaction, from poorly integrated verbal and non-verbal communication to abnormalities in eye contact, body language, gestures, a lack of facial expressions and non-verbal communication.
4. Deficits in developing, maintaining and understanding relationships, from difficulties adjusting behaviour to suit social contexts and difficulties in making friends or the absence of interest in peers.

The definition of Autism Spectrum Disorder also includes limited, repetitive thought patterns, interests, or activities as well as very limited, fixed interests that are abnormally intense or focused. Finally, there is hyper- or hypo-reactivity to sensory stimuli or an unusual interest in sensory aspects of the surrounding.

Asperger is part of the Autism Spectrum Disorder. Individuals with Asperger syndrome often experience a number of problems in daily life: Appropriately recognising and communicating emotions, starting and maintaining friendships or relationships, finding a job, joining a network of colleagues and negotiating problems in the family or in relationships.

It is important to note that people who do not have a diagnosis of ASD, and do not have any mental health problems, may appear robotic as well. The appearance of a limited emotional range and lack of empathy, for instance, are clearly not restricted to individuals who are on the spectrum. Furthermore, there are important cultural differences when it comes to the expression of emotion. It is clearly inappropriate to say that all individuals on the autism spectrum appear to be robotic or lacking in human expressions. As a psychologist who almost exclusively works with clients who are on the spectrum and/or have ADHD, I can assure the reader that individuals with ASD deeply care about others and have no deficit whatsoever in experiencing emotions.

There are also mental health conditions that can give rise to the appearance of being robotic. This includes schizophrenia and certain personality disorders. For instance, a person with schizoid personality disorder may appear to be withdrawn and inexperienced in social communication. This may in turn create the impression of being rigid in expressions and almost robotic. So what it is like for people with symptoms of neuro-developmental disorders and certain mental health conditions, particularly in relation to the notion of the Uncanny Valley?

7.3.1 *Life in the Uncanny Valley*

Goldstein (2020) writes about his experience of being at the high-functioning end of ASD. In Autism Spectrum Disorder, there are also categories for the daily functioning of individuals: (1) Requiring Support, (2) Requiring Substantial Support and (3) Requiring Very Substantial Support. Goldstein (2020) writes that

> the continuum model combined with the perception of a low to high function model creates a bad situation for many working adults, particularly in the technology fields, where greater ability to mask ASD/Asperger's symptoms lead to a greater chance of the *Curse of High Functioning.*

Goldstein (2020) continues to write

> The Curse of High Functioning is the combination of the popular view of function going from low to high with the mistake of believing function is measured against how close to an average neurotypical you appear. Most would consider a person with ASD to be lower functioning if they have communication or social interaction challenges. Because of that, the expectation for the person is lowered and greater allowance is given for odd behaviour. This occurs even if the person is an expert in a technical area. Conversely, if the person is able to display abilities in communication and social interaction at a typical level, they are assumed to be Neurotypical. This is where the Curse of the low, high ASD model appears. Being seen as neurotypical shifts the standards us high functioning autistics have to meet from being challenging to impossible.

It appears that the perception of someone as not communicating well, and in particular lacking in emotional expression, gives license for abuse to some. Throughout human history the worst abuses happened when a group of people was labelled as subhuman or lacking in faith. Individuals with ASD and those who support people on the spectrum often ask the question who is required to make adjustments. The person on the spectrum (Autism Cure: Lets prevent ASD or find a therapy vs. Autism Rights: Society has to adjust to better accomodate individuals on the spectrum).

Appendix C.4 includes a fictional example to illustrate issues in relationships in which one partner has Aspergers. Appendix C includes statements from autistic individuals in the workplace as well, which centre around these individuals' experiences of having difficulties with interpreting other people's perspectives.

7.3.2 Theory of Mind, Autism and Artificial Superintelligence

In the case illustration in Appendix C, Frank clearly did not understand the emotional and mental state of his wife Jill. "Theory of Mind" (TOM) refers to the ability to interpret another person's believes, intentions and emotions (Shvo et al. 2020). This is literally about the appropriate understanding of what another person thinks and feels at any given moment in time. The ability to judge another person's mental state appropriately is a cornerstone of social communication. Only a person who has an accurate theory of mind of others can teach or give explanations that are comprehensible and useful. A deficit in the understanding of the intentions, motives and emotions results in awkward communications or a failure to relate to others. It is commonly believed that the ability to judge the mental states in others is impaired in autism spectrum disorder (Brewer et al. 2017).

With regard to AI, any form of advanced superintelligence must have a highly accurate theory of the mind of humans. If this is not the case, all attempts to communicate will be awkward and unpersonal. Explanations given by the machine would be difficult to understand as well. Yet it is theoretically possible that an advanced form of AI has a superior theory of mind capability compared to humans. In other words, the artificial intelligence might be able to understand humans better than the average person. People who train animals, for instance, can have a better theory of mind than the animal itself; a dog trainer may use a systematic pattern of positive reinforcements to achieve a certain behavioural outcome. In a similar vein, a superior theory-of-mind capability would truly create an artificial superintelligence, since TOM is more important for interaction with humans than pure reasoning and learning. Hence, if a machine has superior reasoning and learning ability, but no appropriate understanding of the mental state of humans, the machine is likely in the uncanny valley: it creates a perception of being robotic, and as a result, may easily be rejected by humans.

There is an obvious contradiction here: If theory-of-mind is so important for communication, why are chatbots successful even though most do not have a TOM of the user? Furthermore, what is the difference between having an accurate TOM and pretending to have one? Currently, the vast majority of interactions with chatbots are goal-directed. For instance, the user wants to have information about a product or wants to complete a transaction. Chatbots are not general artificial intelligence systems. Most conversations are limited to a certain domain and aim at the completion of a limited set of tasks. Furthermore, in many applications, human providers can take over when the chatbot fails to provide a successful service.

There is no doubt that the functionality of chatbots would improve significantly with extended belief systems that include TOM models. However, for many applications this is simply not necessary since chatbots are typically not involved in complex interactions such as negotiating the ups and downs of human relationships. Nevertheless, some of the awkwardness of chatbots can be attributed to the absence of theory of mind models.

There are strategies that can be used to overcome the limitations of a conversational agent without a theory-of-mind model. In the movie "Her", an invisible chatbot with a speech-only interface seduces a user. We do not know if the chatbot in this movie has a TOM, however, it is reasonable to assume that this is indeed the case since the AI system accurately judges the emotional involvement of the user. The key aspect of the movie is that an emotionally vulnerable user interacts with a voice user interface that has all the features to sound seductive and thereby meeting hidden needs and desires. From the viewpoint of user, this is the key aspect and so dominant that minor shortcomings in dialogue and interactions are not noticed. In other words, the *soliciting power* of the chatbot is such that minor errors and failures are disregared.

7.4 Abuse of Chatbots

Conversational agents support users in a number of different ways, including task completion. Yet, studies have shown that a significant part of interactions between humans and conversational agents include abusive language on the part of the user (Kunze 2019). There is a concern that human-chatbot dialogues provide a training ground for verbal abuse and that there can be a transfer to real life interactions between humans. Hence, it is important to understand the nature of these abusive interactions and how conversational agents contribute to negative responses by users. It is useful to have a classification of abusive behaviours by humans and chatbot strategies to mitigate these behaviours.

A person who verbally abuses a software program may think that no harm is done since the recipient is just an information processing machine. Currently, however, harm is indeed done; not to the software of the chatbot but to humans, because the interaction logs of chatbots are monitored by humans for quality assurance.

Certain emotions can enable aggression in interactions while other feelings deter negative behaviours. For instance, guilt and shame are emotions that inhibit verbal aggression. Someone who feels guilty is more prone to apologise then to abuse. Chin and Li (2019) studied the verbal abuse of conversational agents. In their definition,

"Verbal abuse is a hostile form of communication ill-intended to harm the other person" (Chin and Li 2019, p. 1). Three verbal abuse types (Insult, Threat, Swearing) and three response styles (Avoidance, Empathy, Counter attacking) were considered. Chin and Li (2019) studied whether a conversational agent's response style mitigates the aggressive behaviours of users. Sixty-four subjects were assigned to abuse type conditions and interacted with three conversational agents. The results show that, regardless of abuse type, the agent's response style has a significant effect on user emotions. Users were less angry and more guilty when communicating with the empathetic agent than the other two agents that used avoidance or counter attack. Based on their results, Chin and Lin (2019, p. 6) arrived at recommendations for chatbot design:

> First, when users verbally abuse an agent, it is necessary for the agent to ask users about the real intention of their statements, rather than responding to it humorously or providing users with a related search result. Understanding the intent of users and providing a contextual response may allow users to perceive the chatbot as capable, helpful, and enjoyable. Second, if users express negative feelings toward a chatbot, with the intent of abuse, the chatbot should ask users what features of the chatbot make them upset or what situation irritates the user and show a willingness to solve the problem. Our experiment results show that most users have positively assessed the chatbot's attitude of reflecting on its mistakes and asking for user feedback. The chatbot's empathetic attitude toward users' angry feelings would contribute to making itself look less mechanical while reducing users' verbal abuse.

7.4.1 Porn Filters in Chatbots

As illustrated above, one of the first tasks for any chatbot developer is to design a program that responds to verbal abuse. The obvious problem is that poor language can occur in many different forms and can involve many different words and expressions. It is nearly impossible to anticipate all forms of insult and inappropriate language. Frequently, the insult comes with spelling errors. The abuser may be agitated which results in spelling and syntactic errors. In other cases, a speech recognition program may not transcribe the verbal input accurately. It is also possible that the verbal abuse is expressed in a language other than that of the chatbot.

All of the above represent problems for chatbots without full language understanding capability. As a result, verbal abuse may not be recognised in every single instance. An additional challenge is that the bot may be freely available on the internet and this includes cultures with very different norms to the belief system of the designer. For example, there are obvious differences between Islamic and Western cultures when it comes to imagery and verbal expressions. The challenge for the chatbot designer is to develop a program that is both engaging and entertaining but not inappropriate or abusive in any cultural context. One of the solutions is to completely disallow any conversation about sexual topics which is possible in a business and technology context but very difficult in the field of psychology since many mental health problems relate to sexual issues (see Chapter 2).

7.5 Robots for Autism

The use of robots to teach autistic children social skills sounds completely counter-intuitive at first. Natural language and communication skills define us as humans, yet language and communication are areas of relative weakness for machines. Even humanoid robots in the year 2020 are firmly in the uncanny valley and are lacking features that are commonly considered human.

Yet decades of research and development have been devoted to the development of humanoid robots that assist autistic individuals. Specifically, the machines are used for behavioural skill and communication training. The reason is that robots are simply perfect for this task for the following reasons.

Robots are patient and never get tired. Robots cannot get injured in the human sense. Most importantly, robots are good at anything that is repetitive, and repetition is certainly something autistic individuals appreciate and require. At a point when the most caring and emphatic support person of a low functioning autistic individual gets tired because of exhaustion and frustration, a robot simply keeps going. Robots automatically attract attention and this means a human learner will constantly return to the teacher, either the machine itself or another human close by. Robots can be used for fun and games in ways people cannot, so it is very easy to design a learning programme by use of robots that is both instructive and entertaining.

Robots are perfect at teaching behavioural skills by use of methods such as Discrete Trial Training (DTT). DTT breaks down tasks into simpler components that are taught one by one (Steege et al. 2007). It involves four stages: (1) presentation of the discriminative stimulus, (2) performance of the targeted response or an approximation of it, (3) reinforcement that is not directly related to the behavior such as a sticker and (4) inter-trial interval and recording of data.

By way of example, PABI is a penguin-like humanoid robot built to assist in DTT (Moussawi and Diederich 2018). PABI's eyes have cameras, there is a screen in the body plus a motion controller and it is possible to connect a tablet computer. This robot can assess attention level, gestures, gaze direction and affect. The child responds to stimuli presented by the robot on the tablet and PABI then shows behaviours and provides reinforcers or corrections.

Compared to mobile apps, which can consume the attention of the user to the point of being completely absorbed by the device, which in turn, excludes interactions with other people, the presence of a robot actually enhances and stimulates communication between those present (e.g. an autistic individual and a caregiver). With a robot in a room, it is hardly necessary to look for a topic for conversation. When there is a robot around, you simply need to talk about it, and this includes autistic children doing the talking and interacting physically.

When machines operate in an environment that includes people who may attack the device, robots must be protected from failing as a result of physical abuse. As a matter of fact, robots must continue to function, even when damaged, without harming humans. In a previous research project in which I was involved, a lot of efforts went into an ethical clearance application for a robotic device for relatively small, autistic children. This device had to endure physical force and had to continue to function in wet environments without posing a threat to the children. In this case, the machine was designed to tolerate human abuse.

References

Brewer, N., Young, R.L., Barnett, E.: Measuring theory of mind in adults with autism spectrum disorder. J. Autism Dev. Disord. **47**, 1927–1941 (2017)

Bromwich, J.E.: Why do we hurt robots? New York Times, 19 Jan 2019

Chin, H., Li, M.Y.: Should an agent be ignoring it? A study of verbal abuse types and conversational agents' response styles. In: CHI 2019, 4–9 May 2019. Glasgow, Scotland, UK

Darling, K.: Children beating up robot inspires new escape maneuver system. In: IEEE Spectrum. https://spectrum.ieee.org/automaton/robotics/artificial-intelligence/children-beating-up-robot (6 Aug 2015)

Goldstein, T.: Planet neurodivergent. https://planetnd.market/articles-and-videos/autisms-uncanny-valley-the-curse-of-being-high-functioninglow-support (24 Sept 2020)

Kunze, L.: Chatting with machines: Strange things 60 billion bots logs say about human nature. https://youtu.be/KSJ0DMQbN-s (2019)

Mori, M.: The uncanny valley. IEEE Robotics and Automation Magazine, pp. 98–100 (2012)

Moussawi, N.A., Diederich, J.: Technological Interventions for Autism Spectrum Disorder (ASD). http://doi.org/10.17605/OSF.IO/MDQWH (9 Mar 2018)

Shvo, M., Klassen, T.Q., McIlraith, S.A.: Towards the Role of Theory of Mind in Explanation. Chapter. https://doi.org//10.1007/978-3-030-51924-7_5 (July 2020)

Steege, M.W., Mace, F.C., Perry, L., Longenecker, H.: Applied behavior analysis: Beyond discrete trial teaching. Psychol. Sch. **44**(1), 91–99 (2007). https://doi.org/10.1002/pits.20208

Wang, S., Cheong, Y.F., Dilks, D.D., et al.: The uncanny valley phenomenon and the temporal dynamics of face animacy perception. Perception. https://doi.org/10.1177/0301006620952611 (9 Sept 2020)

Appendices

Appendix A: Knowledge Acquisition and Automated Psychotherapy

Knowledge Acquisition and Kriton Speech

The conversational agent JoBot™ introduced in Chapter 2 utilises the Kriton Speech system. Kriton Speech integrates cognitive computing and psychological methods to acquire knowledge for artificial intelligence systems and to apply this knowledge for computational diagnosis and therapy. The original Kriton system was developed as a general knowledge acquisition system and has been adopted for psychological applications (Diederich 1987a, d, 1988). As a knowledge acquisition system, Kriton Speech elicits domain-specific knowledge and as an application, it applies this knowledge for diagnosis and therapy. Kriton Speech exclusively uses a voice user interface (VUI). The system uses knowledge bases such as upper-level ontologies as well as domain-specific knowledge and psychological heuristics (1) to interview the user in order to obtain knowledge, and based on this knowledge, (2) perform tasks such as psychological assessment and therapy. The system allows free verbal responses and has an explanation capability, i.e. provides feedback about why certain questions are asked. Speech and language are analysed to determine, with high accuracy, if there is a mental health problem such as depression.

Attention economy is a concept in information management that treats human attention as a limited commodity. Herbert A. Simon (1971) is considered to be the first person having introduced the concept of attention in economics. Simon (1971) outlined that in an information rich world there is a limitation of the resource that consumes information: human attention. Almost half a century later, we are now at the point where there are so many information sources, so many devices and so many distractions that human attention is not only limited, it is frequently impaired.

Kriton Speech addresses both problems: the limitation of human attention and the challenge of growing knowledge. Firstly, Kriton Speech uses speech only to interact

J. Diederich, *The Psychology of Artificial Superintelligence*,
Cognitive Systems Monographs 42,
https://doi.org/10.1007/978-3-030-71842-8

with the user and therefore does not require any visual attention. This enables the user to direct attention to other information sources and to perform physical actions while using the program. No learning is required, any user can interact with Kriton Speech immediately. It is possible to drive a car while talking to Kriton Speech; to go for a walk or to perform other activities.

Kriton Speech uses upper-level ontologies as well as domain-specific knowledge to interview the user. The employment and use of already acquired knowledge has major advantages: (1) The interview aims at completing and refining existing knowledge in addition to the acquisition of new concepts and rules. (2) On the application side, the system utilises up-to-date clinical knowledge obtained from practising mental health experts to conduct psychological assessment and therapy.
Below please find a sample dialogue with a user:

Kriton: What would you like to work on?
User: Mental health and related issues.
Kriton: Name three concepts that are important in the context of mental health.
User: Asperger, ADHD and Anxiety.
Kriton: Are these different concepts?
User: Yes.
Kriton: Do you want to consider Social Communication Disorder as well?
User: No, but I want to include ADD.
Kriton: What do Asperger and ADHD have in common that Anxiety does not have?
User: These are neurodevelopmental problems.
Kriton: What do Anxiety and Asperger have in common that ADHD does not have?
User: Social Discomfort.
Kriton: Do you want to accept DSM-5 concept definitions?
User: Yes.

As a result of the interview, Kriton Speech builds and verifies ontologies and rule based systems. The output is a Web Ontology Language (OWL) file that can be edited by the use of standard ontology editors. Information obtained from clients is stored in an XML format to be used by eHealth systems (Fig. A.1).

Amy: An Artificial Intelligence ADHD Coach

Amy is an artificial intelligence (AI) system that provides psychological assistance simply by talking to a client. When linked to background information in the form of ontologies as shown above, Amy has extensive medical and psychological knowledge and the built-in capacity to acquire additional knowledge from mental health experts. In turn, this knowledge is used for psycho-education. Amy's primary mode of communication is a simple and convenient conversation, however, Amy also provides videos, images and text to educate users about mental health issues.

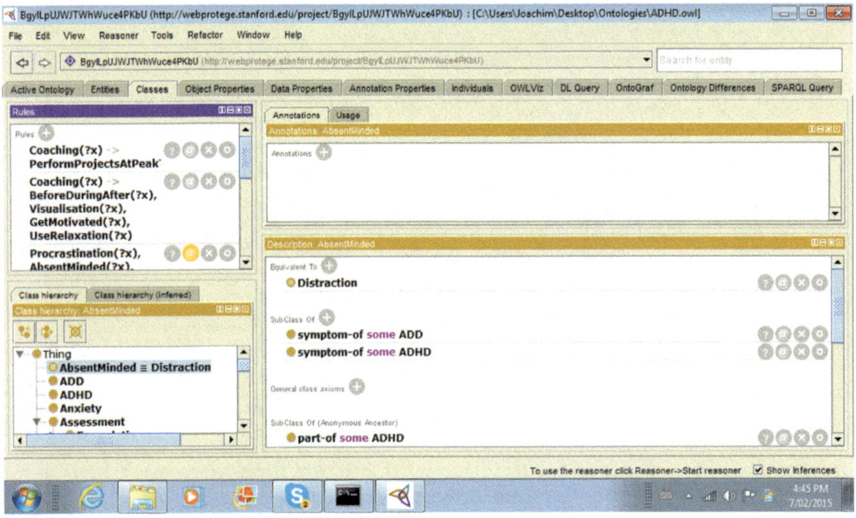

Fig. A.1 An ontology for ADHD coaching

Amy operates in a number of different modes. If the user is inclined to have a free conversation about any topic or to obtain information about ADHD, Amy is simply a chatbot with a very strong protection against inappropriate content. With machine learning enabled, Amy detects mood changes in speech and text including those pointing to depression and the system can modify the interaction accordingly. While Amy has the ability to learn, the dialogue structure is rule-based and does not change. As a result, Amy is always socially and culturally appropriate. Amy frequently refers to online resources to illustrate recommendations, including the use of relaxation techniques (Fig. A.2).

References

Diederich, J., Ruhmann, I., May, M., KRITON: A knowledge acquisition tool for expert systems. Int. J. Man-Machine Stud. **26**, 29–40 (1987)

Diederich, J., Ruhmann, I., May, M., KRITON: A knowledge acquisition tool for expert systems. In: Boose, J.H., Gaines, B. (eds.) Knowledge Acquisition Tools for Expert Systems. Knowledge-Based Systems 2, pp. 83–94. Academic Press, London etc. (1988)

Diederich, J., Techniken des Wissenserwerbs (Knowledge Acquisition Techniques). In: Christaller, T. (ed.) 5. Künstliche Intelligenz Frühjahrsschule, pp. 295–335. Springer Verlag 1988 = Springer Informatik Fachberichte 202, Berlin (1987)

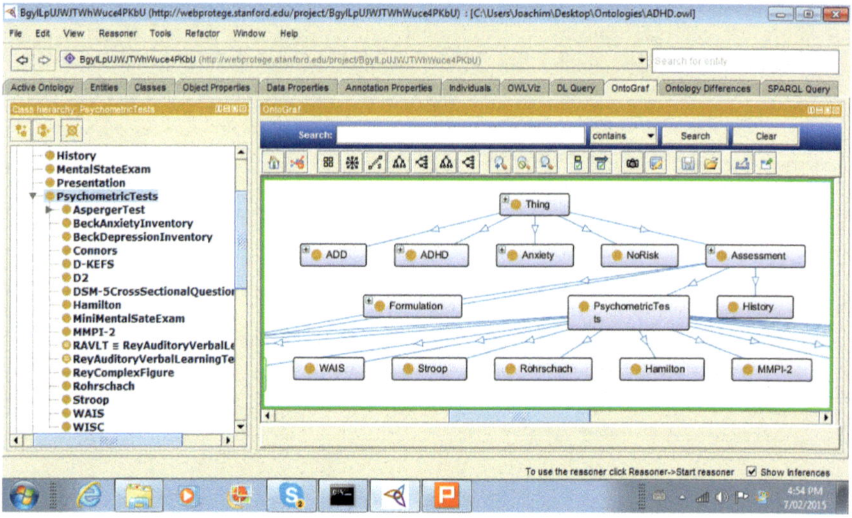

Fig. A.2 An ontology for ADHD coaching

Simon, H.A.: Designing organizations for an information-rich world. In: Martin Greenberger, Computers, Communication, and the Public Interest, pp. 40–41. The Johns Hopkins Press, Baltimore, MD (1971)

Appendix B: Speech, Language and Mental Health

Language, Depression and Schizophrenia

There are speech and language differences between individuals affected by schizophrenia, depression and those free of symptoms. Firstly, several studies demonstrate that schizophrenic patients communicate less compared to control subjects (Andreasen and Grove 1986; Diederich 2017). Qualitatively, schizophrenic speech and language disorders includes two features which are commonly referred to as 'poverty of speech' (lack of speech or reduced vocabulary) and 'poverty of content' (lack of variety). Schizophrenic speech and language disorder persists longitudinally, occurring in about 80–90% of patients during the acute stages of hospitalisation and prevailing in about 39% of the patients during the post-acute stages (Andreasen and Grove 1986).

Since speech and language disorders are associated with cognitive impairments, verbal communication is regarded as a reliable clinical marker of a psychotic disorder as it can be directly observed. Speech and language disorders (SLDs) are also considered an early vulnerability indicator of schizophrenia since their presence in children and adolescents is a significant predictor of adult psychosis. Unlike positive speech and language disorder (e.g. derailment, incoherence), it is the negative type (poverty of speech and content) that is indicative of a poor prognosis (Andreasen and Grove 1986). Therefore, a correct assessment of SLD should be part of clinical assessment (including the determination of the prognosis).

Smirnova et al. (2018) observed that

> deviations from typical word use have been previously reported in clinical depression, but language patterns of mild depression (MD), as distinct from normal sadness (NS) and euthymic state, are unknown.

Smirnova et al. (2018) studied 402 written reports from 124 Russian-speaking patients and 77 healthy controls (HC), including 35 cases of normal sadness using hand-coding procedures. Smirnova et al. (2018) reported that in mild depression, compared to individuals without symptoms,

- written responses were longer,
- a descriptive rather than analytic style was used,
- there are signs of spoken and figurative language,
- single-clause sentences were more frequent than complex sentences,
- there was unusual word order,
- there was increased use of personal and indefinite pronouns, and
- more verb use in continuous/imperfective and past tenses, indicating that the individual is occupied with past events.

In natural sadness compared with controls, Smirnova et al. (2018) found more lexical repetitions, omission of words, and verbs in continuous and present tenses.

Mental health practitioners, consciously and unconsciously, analyse the language of their patients, identify patterns and use this information for clinical assessment, or classification. Machine learning techniques can be applied to analyse data, including speech and language.

Support Vector Machine Learning

Support vector machines are an alternative to neural networks as tools for solving pattern recognition problems. SVMs have a major advantage over neural networks in that they formulate the learning problem as a quadratic optimisation problem whose error surface is free of local minima and has a unique global optimum. SVMs are well-suited to learning classification and regression tasks. SVMs have also been used for ranking problems, e.g. a ranking with regard to the severity of a disorder. Furthermore, SVMs have been utilised in a wide variety of tasks, including text and image classification as well as bio-medical applications.

SVMs are based on simple ideas and provide a clear intuition of what learning from examples is all about. More importantly, they also show high performance in practical applications. SVMs correspond to a linear method in a very high dimensional feature space that is non-linearly related to the input space. Even though SVMs implement a linear algorithm in a high dimensional feature space, in practice they do not involve any computations in that high dimensional space. By use of kernels, all necessary computations are performed directly in input space. Data vectors nearest to the separating hyperplane in the transformed space are called support vectors. Classification as well as regression can be learned by SVMs.

In binary classification tasks, the margin between the two classes is maximised in order to find the best possible separation. An SVM algorithm finds an optimal decision boundary in the multi-dimensional feature space by finding a hyperplane, which has maximum distance from prototypical samples, called "support vectors". The learned hyperplane is then used as a decision boundary of the SVM classifier.

To allow SVM learning in systems like JoBot™, speech samples must be transcribed, the language input is then segmented into words, each word is stemmed and functional words (such as "to" and "the") are removed. Words are ranked according to frequency and words with less than two occurrences are removed. The selected words are then used as attributes, and thus each text is represented as a vector of the frequencies of words.

In more detail, the input features are normalised frequencies of words. Words in the whole corpora are ranked and infrequent words are discarded. The normalised frequency (normalised to the length of input) of each word in the ranked list of words are then used as input features of each document. That is, an input feature $x = (v_1,\ldots,v_d) \in \Re^d$ is d-dimensional vector and each feature value v_i is a normalised frequency of a word:

$$v_i = \text{Frequency of the } i - \text{th Word/The Length of Document}$$

where the length of document is the L2-Norm of word frequencies in a document. The task is then to find a function $f{:}x \rightarrow D$ by use of SVM learning that maps the input feature vector $x \in \Re^d$ to a class label $D \in \{+1, -1\}$.

Mood Analysis in JoBot™

JoBot™ offers a number of independent forms of language analysis to determine the mood of the user, the machine learning method described above among them. The analyses operate at the dialogue and word level. First of all, JoBot™ allows any form of dialogue and any form of input. The user is free to say whatever s/he wants and is only prevented from conducting an inappropriate sexual conversation. If inputs are detected that indicate a cognitive dysfunction, thought disorder or negative mood, then a structured dialogue is entered. During a structured dialogue, the user is still free to say whatever s/he wants, however, a number of responses are suggested which allow to gather additional information on past psychological issues or the current mental state.

Structured dialogues either lead to the presentation of information on mental health (psychoeducation) or the description of a psychological intervention. If the user offers information during a structured dialogue that point to previous mental health problems or current mood issues then a number of independent scores are updated (in most cases, values are increased). Receiving information on mental health problems or a description of a psychological intervention by the user decreases these scores. Scores may return to zero if the user is fully engaged with psychological therapies such as CBT.

Word level mood analysis is partially motivated by the work of Rude et al. (2004). This approach in turn is based on Beck's (1967) cognitive theory of depression which holds that individuals may have knowledge structures, so-called depressive schemas, which result in a very negative outlook to the world. Depressive schemas lead to automatic thoughts and unhelpful thinking styles which are viewed as a core cognitive component of depression. In addition, it has been observed that depressed individuals very much focus on themselves and that this in turn affects their language. The self-focus can reinforce negative emotions, self-blame and grief.

Penebaker, Francis and Booth (2001, in Rude et al. 2004) used a computerised text analysis program to compare poems from nine authors who later committed suicide. The Linguistic Inquiry and Word Count (LIWC) program helped to determine that poets who later committed suicide use more first person, singular words such as "I, me and my" and fewer words pointing to social groups (we, us, our). It is not clear if this observation is due to suicidality or depression in the selected group of poets.

Rude et al. (2004) studied psychology undergraduates from an introductory class who completed a questionnaire and essays. Subjects were assigned to an experimental and control group based on the result of self-report depression questionnaires. The linguistic analysis was performed by the LIWC. Depressed students revealed both self-focus and -preoccupation, however, social disengagement was not observed in this study.

Currently, JoBot^{TM} is analysing the use of words representing self-reference such as "I, me and my", in addition to the consideration of 4551 negative emotion words as well as 19 "absolutist" words in the English language (absolutist words such as "all" or "nothing" are an indication of "black and white" thinking, a cognitive dysfunction in depression, see Al Mosaiwi and Johnstone (2018). The occurrence of any word indicating self-reference or any of the negative emotion words increases one of several scores; the mood score. The mood score indicates the level of self reference in the language of the user in addition to the occurrence of negative statements.

JoBot^{TM} makes reference to the mood score in the context of another variable, the mood history. The mood history score is based on the occurrence of complete statements indicating depression. For example, a statement such as "I am very sad" would increase the mood history score since the current mental state is described as sadness. All mood scores are reset if the client engages with psychological therapy or psychoeducation (Table B.1).

Table B.1 Appendix B: Mood analysis in JoBot™

Text analysis level	Method	Objective
Dialogue	Structured graph analysis	Analyse self-reported depression statements
Full text	Machine learning	Classification and ranking wrt depression
Word	Count depression words	Determine negativity of language

References

Al Mosaiwi, M., Johnstone, T.: In an absolute state: Elevated use of absolutist words is a marker specific to anxiety, depression, and suicidal ideation. Clin. Psychol. Sci. **6**(4), 529–542 (2018)

Andreasen, N.C., Grove, W.M.: Thought, language, and communication in schizophrenia: Diagnosis and prognosis. Schizophr. Bull. **12**, 48–359 (1986)

Beck, A.T.: Depression: Clinical, Experimental, and Theoretical Aspects. Harper & Row, New York (1967)

Diederich, J.: Simulation schizophrener Sprache (Modelling Schizophrenic Language). Springer Verlag, Berlin, Heidelberg, New York (2017). Reprint. ISBN: 978-3-531-19154-6

Rude, S., Gortner, E.M., Pennebaker, J.: Language use of depressed and depression-vulnerable college students. Cognit. Emot. **18**(8), 1121–1133 (2004)

Smirnova, D., Cumming, P., Sloeva, E., Kuvshinova, N., Romanov, D., Nosachev, G.: Language patterns discriminate mild depression from normal sadness and euthymic state. Front. Psychiatr. **9**, 105 (2018)

Appendix C: Case Illustrations

In order to illustrate the utility of the Psychotherapy as Heuristic Search framework, a number of case studies are summarised below and the process of psychotherapy is translated into the conceptual structure introduced in Chapter 2.

Little Albert

Little Albert is among the most famous case studies in psychology. Watson and colleagues used classical conditioning to demonstrate how a form of "neurosis" can develop in a child (Trull and Prinstein 2013). A young boy who was not afraid of any furry animals was presented with a white rat and while he was playing with the animal, a loud noise was made behind him. After few trials, Albert developed a fearful response and this generalised to other furry objects (the ethics of this study and the experimental set-up have been severely criticised for decades).

Jones (1924; in Trull and Prinstein 2013) demonstrated how fear can be removed in an independent study with a three-year old boy. Peter was afraid of rabbits and similar animals. In order to treat this form of anxiety, the researchers brought a caged rabbit closer and closer to the boy as he was eating. As a result, the dreaded object became associated with food and after some period of time, Peter's fear of rabbits disappeared. It was important that the researchers assessed the fear of rabbits to make sure it was not so intense that the child would develop an aversion to food. This tightly controlled treatment was successful.

Jones (1924; in Trull and Prinstein 2013) had a clearly defined starting point for her psychological therapy: Peter and his fear of rabbits. In addition, there are methods which are now part of behaviour therapy to eradicate the fear (in this framework, these are the means to move from one sub-goal to the next). The starting point was Peter with a set of symptoms (fear of a rabbit). The goal state was the client free of any symptoms. Step-by-step, the researchers brought the cage with the rabbit closer and closer to Peter. By doing so, the investigators moved from the start state to intermediate states and finally to the goal state (a client relieved of this particular fear). Please note that while the individual actions (moving the cage) are almost identical, the intermediate *problem solving states* in this particular form of therapy differ. Each state has a different probability to develop an aversion to food! Assessment was required to make sure that this food aversion did not happen. Clearly, an intermediate state with a higher probability of an aversion to food is less desirable than a state that brings the client closer to a freedom of symptoms with low probability of developing negative side effects. Peter's treatment can be clearly defined within the problem solving framework.

Little Hans

This very famous case study by Sigmund Freud is about childhood anxiety as well. Hans' (not his real name) father belonged to Freud's circle of followers in Vienna and he noted that Hans developed a fear of horses. The father wrote to Freud (Strachey 1909):

> No doubt the ground was prepared by sexual over-excitation due to his mother's tenderness; but I am not able to specify the actual exciting cause. He is afraid a horse will bite him in the street, and this fear seems somehow to be connected with his having been frightened by a large penis.

Alternative explanations for the fear of horses have been offered, including the observation that five-year old Hans may have been frightened by a collapsing horse on the street.

Freud had a sexual interpretation of the fear of horses broadly based on the concept of the Oedipus complex. With this understanding, Freud started the process of psychoanalysis. There was no follow-up, however, fourteen years later, Hans reported to be fine and well-adjusted without memory of his childhood phobia.

Freud stated that psychoanalysis is a therapeutic measure and not a scientific investigation (Wolpe and Rachman 1960, p. 145). It appears that in psychoanalysis, the interpretations provide the means to achieve goals such as freedom of symptoms. In this case, no fear of horses.

A Fictional Case Study: Maryam

Maryam was referred by her mother Hibba as she is exhibiting anger and conduct problems at home. Hibba perceived Maryam to be a challenging child since birth. At the age of two, parenting advice was sought from a psychologist since Hibba was overwhelmed by the experience of having a child and this resulted in arguments with her partner Omar as well. Both parents found it difficult to respond positively to Maryam and aversive interaction patterns emerged, between both parents we well as the parent-child system. The aversive behaviour of her parents provided modelling opportunities for Maryam and her self-esteem deteriorated in recent years since her academic performance is below average. There are significant protective factors that help to prevent a deteriation of mental health: Maryam lives in a stable home environment; she is physically healthy, both parents are employed and have sought help for Maryam in the past. In addition, no conduct problems at school have been reported. Maryam's behaviour should improve if the family commits to a psychological treatment based on cognitive behaviour therapy (CBT) and improvements in the family system.

The following long-term goals have been identified:

1. To reduce Maryam's anxiety, in particular separation anxiety, as well as feelings of panic due to separation fears.
2. To reduce periods of low and irritable mood as well as episodes of anger.

The long-term goals will be achieved if there is (1) a significant reduction in anxiety and depression scores (self- and parent assessment), and (2) a reduction in observer-rated scores for mood problems. This goal should be achieved after 4–6 months (8–12 sessions).

In addition, the following mid-term goals have been identified: To increase periods away from mother Hibba without expressed fear or anger as well as sleeping in Maryam's own bedroom on the other side of the house. The mid-term goal will be achieved if (1) Maryam attends day-time play-dates on her own twice a week and (2) sleeps in her own bedroom several times a week (after 4–8 weeks). Progress towards these goals will be monitored during the first 2–8 sessions.

Finally, here are the short-term goals for therapy: Maryam to identify negative thinking patterns related to separation anxiety and replace these with more positive thoughts. This will be measured by Maryam expressing these skills during sessions and completing homework assignments for two consecutive sessions. Maryam should achieve these goals within 2–4 weeks.

After identifying clear goals for psychotherapy, the means for transitioning from one state to another are identified. In this case, CBT and family therapy. Barrett et al. (2001, pp. 135/6) observe that CBT alone as well as CBT plus family management/treatment showed greater improvement on a variety of measures at a 12-month follow-up when compared with a group without treatment. More specifically, "clients that received family training also showed significantly greater improvement than the CBT-only group on a number of measures" (Barrett et al. 2001, p. 136).

After the identification of goals and sets of psychological interventions that allow to move from one intermediate state to another, the sessions can proceed as follows:

(a) First session (with Maryam and her mother): Learning to beat anxiety. The foundations of CBT are explained to Maryam and Hibba.
(b) Second session (with Maryam): Introduction of the "Hot Diary" (a booklet with negative, problematic thoughts). Several sample situations are discussed.
(c) Third session with the entire family: Explain attachment theory and "circle of security". Also, explain the importance of a "secure family base".
(d) Fourth session (with Maryam). Introduction to controlled breathing exercises. Discussion of the Hot Diary and fears in relation to sleeping in Maryam's own bed.

In this case study, goal decomposition is used to determine the desired final outcome but also to identify sub goals that constitute progress in psychotherapy. For benchmarking purposes, it is clearly stated which goals should have been reached after a certain number of sessions. Interventions from cognitive behaviour therapy as well as family therapy are the actions that allow to traverse problem space, that is to move from one sub-goal to the next until the final outcome is achieved. In addition, the criteria for achieving goal state are very clear: scores in psychological tests that measure child anxiety.

A Fictional Case Study for Asperger Relationships: Frank and Jill

Frank and Jill have been married for almost 20 years. Jill has been unhappy with the relationship for a long time and she saw a number of psychologists who suggested her husband Frank may have Asperger's. Frank does not believe there is anything wrong with him, however, after frequent request by Jill, he went to see a clinical psychologist for an assessment. It confirmed high functioning autism spectrum disorder without an impairment of speech and language or any form of intellectual disability. As a matter of fact, Frank is a very successful engineer who works in his own business.

Motivated by the recent assessment, Jill and Frank saw a relationship counsellor for help. During the first session, Jill stated that she does not feel loved even though Frank tells her so frequently. Jill complains that she feels invisible to Frank and that he never acknowledges her presence. In her opinion, Frank works all the time and is not really interested in sharing his thoughts and emotions. In contrast, Frank told the counselor that he deeply loves Jill, that she is the love of his life and that he would do anything to save the marriage.

Jill reported that she noted something was not quite right after the first honeymoon period of the relationship. It was as if all the love had left Frank. Jill decided to do everything possible to help Frank, so she searched the Internet for solutions. A colleague at work told her about Aspergers and once she got all the information, Jill convinced herself that her partner must be on the spectrum.

Jill told the relationship counsellor that she did everything possible to save a loving relationship with Frank. Helping Frank became her project. If she would use all her love and energy to teach Frank how to communicate and express love, everything would be good. But Frank did not change in her eyes and the years drained her energy and optimism. Despite her best efforts, "Frank remained robotic". As a result, Jill said she became cold and even resentful. Nobody has fought for a relationship more than her and the result is nothing but a loveless, sterile partnership.

With some regret, Jill told the relationship counsellor that she became so resentful that she did a number of things she knew Frank would hate. She took his socks and put them into a drawer where he would not find them. Frank loves to sort things by colour, in particular his clothing. Jill made sure everything in the house was organised in a way completely inconsistent with Frank's preferred colour coding. In complete desperation, she threw Frank's work laptop out of the window and replaced it with a toy robot that she put on his desk at home. Frank was annoyed but did not get the message.

Frank did everything possible to convince the relationship counsellor that he loves his wife and while the relationship may have become more mature over the years, his positive feelings did not change. The couple decided to separate for a few months, mostly because Jill felt emotionally exhausted and drained.

Observations of Individuals with ASD

Rebholz (2012, p. 56) provides statements of employees with ASD in the workplace:

Gina: The hardest thing at my job is that I don't think anybody knows that I put on a really good act. Nobody knows that there's anything wrong, I think. Then when something goes wrong, they don't take into account that you're working 10 times harder. [I would like it if my peers could] respect that and maybe empathize little more about how much more effort we have to put in for things that would come really, really easily to other people.

Ian: The other thing, too, is that I don't like—believe it or not—face-to-face meetings, and that's difficult for a sales person to say. But my hands sweat; I shake; I'm super-nervous when I go into a face to face meeting. But if I stay factual and I present, especially if I present over the phone or mail in a presentation, I'm really successful at it. Where I struggle is when I'm face to face with someone, and then there's a social component (Rebholz 2012, p. 56).

Rebholz (2012, p. 64) continues with this statement

Hank: I don't like working with other people. I mean, it's not that I don't like working with other people, but I don't like sitting with other people all day long. Cubicles are really distracting for me, quite honestly. And the fluorescent lights. It's not that I have an issue with the fluorescent lights necessarily, I just don't like 'em. The people talking, and I just don't like it. I'd like a place where I'm able to be in either an office … with just the people that I'm working with, or I can telecommute, or one of those things, because I can look at the wall, and I'm going to get distracted, because I get distracted pretty easily. Then I'm going to Facebook or something. Yeah. And I've never brought that up to anybody, it's just what happens. I try to make sure that I get what I need without sounding like I need it, if that makes sense.

A Fictional Case Study: Martin

Martin is a 35-year-old university student currently enrolled in an education degree with the aim of becoming a secondary school teacher. Martin used to work on construction sites until he decided to finish his high school diploma to get a university degree. At the commencement of therapy, Martin needed to complete two courses in physical education and physics for the award of his degree. As part of the final examinations, it is a requirement to reflect on the relationship between the trainee teacher and the pupils, something Martin finds very awkward and difficult.

The assessment at the start of therapy points to an autism spectrum disorder without an impairment of intellectual functioning or speech and language processing. The assessment also points to ADHD comorbid with the autism spectrum disorder. Martin was referred to a psychiatrist for further assessment and medical advice.

In this scenario, the clinical psychologist completed the assessment and provided documentation for the disability services at Martin's university so that all necessary support for Martin could be set in place. After completion of the assessment and submission of all reports, the provisional psychologist (a trainee psychologist who works under supervision) met with Martin online to teach social communication skills and to introduce Martin to relaxation techniques that may be helpful in situations that cause social stress. The provisional psychologist referred Martin to the conversational agent to get additional information on psychological techniques and to ask questions about these methods.

In this scenario, quality assurance is also performed by the conversational agent. After each session, Martin is referred to interactive questionnaires to obtain feedback on the quality of the online meeting.

References

Barrett, P.M., Duffy, A.L., Dadds, M.R., Rapee, R.M.: Cognitive-behavioral treatment of anxiety disorders in children: Long-term (6-year) follow-up. J. Consul. Clin. Psychol. **69**, 135–141 (2001)

Rebholz, C.H.: Life in the uncanny valley: Workplace issues for knowledge workers on the autism spectrum. PhD Dissertation. Antioch University AURA—Antioch University Repository and Archive (2012)

Strache, J.: Analysis of a phobia in a five-year-old boy. In: The Standard Edition of the Complete Psychological Works of Sigmund Freud, vol. X: Two Case Histories ('Little Hans' and the 'Rat Man'), pp. 1–150 (1909)

Trull, T.J., Prinstein, M.J.: Clinical Psychology, 8th edn. Wadsworth CENGAGE Learning, Belmond CA (2013)

Wolpe, J., Rachman, S.: Psychoanalytic "evidence": A critique based on Freud's case of little Hans. J. Nerv. Mental Dis. **131**, 135–148 (1960)

Printed by Printforce, the Netherlands